Buffers for pH and Metal Ion Control

Chapman and Hall Laboratory
Manuals in Physical Chemistry and Biochemistry

Consulting Editor
ADRIEN ALBERT, Professor Emeritus,
Research School of Chemistry,
Australian National University

In the same series
The Determination of Ionization Constants
Adrien Albert and E. P. Serjeant

Buffers for pH and Metal Ion Control

D. D. Perrin
John Curtin School of Medical Research
Australian National University
Canberra

Boyd Dempsey
Faculty of Military Studies
University of New South Wales
Royal Military College
Duntroon

CHAPMAN AND HALL
London

A HALSTED PRESS BOOK
JOHN WILEY & SONS
New York

First published 1974
by Chapman and Hall Ltd
11 New Fetter Lane, London EC4P 4EE
© 1974 D. D. Perrin and Boyd Dempsey
Typeset by Santype Ltd (Coldtype Division)
Salisbury, Wiltshire
Printed in Great Britain by
Fletcher & Son Ltd, Norwich

Library of Congress Cataloging in Publication Data

Perrin, Douglas Dalzell.
    Buffers for pH and metal ion control.

    Includes bibliographical references.
    1. Buffer solutions. 2. Hydrogen-ion concentration.
3. Metal ions. I. Dempsey, Boyd, joint author. II. Title.
QD56k,P44   541'.34   74-2799
ISBN 0-470-68067-9

Contents

Preface

This book is intended as a practical manual for chemists, biologists and others whose work requires the use of pH or metal-ion buffers. Much information on buffers is scattered throughout the literature and it has been our endeavour to select data and instructions likely to be helpful in the choice of suitable buffer substances and for the preparation of appropriate solutions. For details of pH measurement and the preparation of standard acid and alkali solutions the reader is referred to a companion volume, A. Albert and E. P. Serjeant's *The Determination of Ionization Constants* (1971).

Although the aims of the book are essentially practical, it also deals in some detail with those theoretical aspects considered most helpful to an understanding of buffer applications. We have cast our net widely to include pH buffers for particular purposes and for measurements in non-aqueous and mixed solvent systems. In recent years there has been a significant expansion in the range of available buffers, particularly for biological studies, largely in consequence of the development of many zwitterionic buffers by Good *et al.* (1966). These are described in Chapter 3.

However, there are very many substances that could be, or have been, of use as buffers, and Appendix III lists some of these. Chapter 5 shows how new pH-buffer tables can be constructed from the thermodynamic pK_a values, and some simple computer programmes are included to facilitate the necessary calculations. Tables and worked examples are given for use if a computer is not available.

In view of the importance of metal ion concentrations, particularly in biological work, we have considered it appropriate to include a section on metal-ion buffers. These buffers may also be useful in preparing convenient standards for ion-selective electrodes.

Canberra,　　　　　　　　　　　　　　　　　　D. D. Perrin
August, 1973　　　　　　　　　　　　　　　　Boyd Dempsey

Chapter One

Introduction

1.1 The concept of buffer action

When partly neutralized weak acids or bases are present in aqueous solution the addition of small amounts of strong acid or strong base causes little change in pH. This resistance to change in the free hydrogen ion concentration of a solution was described by Fernbach and Hubert (1900) as 'buffering'. In studies of the enzyme, amylase, they found that a partly neutralized solution of phosphoric acid acted as a 'protection against abrupt changes in acidity or alkalinity: the phosphates behave as a sort of buffer'. Following this observation use was soon made of mixtures of monohydrogen phosphate/dihydrogen phosphate, ammonia/ammonium chloride, acetate/acetic acid, phthalate/phthalic acid, and p-nitrophenolate/p-nitrophenol to obtain solutions which were 'practically unaffected by the presence of traces of (acidic or basic) impurities in the water or salts used, which is far from being the case with very dilute solutions of strong acids and bases' (Fels, 1904).

In Lowry-Brönsted terms, the common feature of these mixtures is the presence of an acid and its conjugate base. This acid-base pair, together with appropriate counter ions, constitutes a buffer substance.

A convenient definition (Van Slyke, 1922) of a buffer is a 'substance which by its presence in solution increases the amount of acid or alkali that must be added to cause unit change in pH.' Addition of 1 ml of 1M HCl to a litre of distilled water, pH 7, lowers the pH to 3. Alternatively, addition of 1 ml of 1M NaOH raises the pH to 11. Much less change (only about 0.02 pH units) is observed if the same amount of acid or alkali is added to a litre of solution, also pH 7, that is 0.05M in imidazolinium hydrochloride and 0.047M in imidazole. Thus imidazole/imidazolinium ion is a good buffer at pH 7.

Another useful property of a buffer, at least within the range pH 4 to 10, is that its pH remains substantially unchanged upon dilution of the solution. The effectiveness of a buffer depends on its buffer capacity (resistance to pH change on addition of acid or alkali), the pH change on dilution, and the effects of adding neutral salts or changing the temperature. These are discussed in Chapter 2.

1.2 Why are buffers needed?

Many biological and chemical systems involve acid-base equilibria and therefore depend critically on the pH of the solution. An example is the extent to which the viability and growth of organisms and tissues depends on the pH of the cell fluids and of the media in which the cells grow (Albert, 1968).

The effectiveness of many chemical separations and the rates of many chemical reactions are governed by the pH of the solution. Buffer solutions offer advantages for controlling reaction conditions and yields in organic syntheses. The objection that this adds material which must later be removed is less serious if volatile buffers are used. A particularly promising field in this respect is biomimetic chemistry (Breslow, 1972) which attempts to imitate natural reactions and enzymic processes as a way to improve the power of organic chemistry.

In analytical and industrial chemistry, adequate pH control may be essential in determining the courses of precipitation reactions and of the electrodeposition of metals. Physico-chemical studies of reaction kinetics and chemical equilibria often require solutions to be maintained at a definite pH value. Buffers are needed for pH standardization and control in the research laboratory, the factory and the medical clinic. For kinetic, equilibrium and physiological studies it is often desirable to make measurements over a controlled range of pH values while, at the same time, maintaining constant ionic strength in the medium.

Much of the success of complexometric methods of chemical analysis depends on the use of buffers to maintain pH constancy so that small changes in free metal ion concentrations can be detected by suitable metallochromic indicators.

1.3 Some naturally occurring buffers

The pH of mammalian blood reflects the state of acid-base balance of the body. It is normally maintained close to a value of 7.38 by the interaction of a series of intricate mechanisms which involve the production and buffering of acid, and its elimination, by the body. The more important of the equilibria comprise inorganic systems, such as $H_2PO_4^-$/ HPO_4^{2-}, CO_2, H_2CO_3/HCO_3^-, and organic acidic and basic groups, particularly of proteins. These include substituted imidazolinium/substituted imidazole (of histidine), substituted phenol/substituted phenolate, -SH/-S$^-$, -COOH/-COO$^-$, -NH$_3^+$/-NH$_2$ and organic phosphorus compounds.

A major part of the buffering capacity of blood is also due to the haemoglobin/oxyhaemoglobin equilibrium and its effect on carbon dioxide transport, but the quantitative treatment of acid-base equilibrium in blood is complicated by the rates of reversible hydration of carbon dioxide (catalysed by carbonic anhydrase) and by the chloride shift from erythrocytes.

Living plant tissue is also buffered, but less closely, the normal pH range in vegetative tissue being 4.0 to 6.2. The main buffers are phosphates, carbonates and organic acids, commonly malic, citric, oxalic, tartaric and some amino acids, with smaller amounts of other organic acids which are intermediates in biochemical cycles.

Open ocean pH values usually lie within the range pH 7.9–8.3. Sillén (1967) showed, on the basis of a multicompartment model, that this constancy was due mainly to equilibria with aluminosilicates, and that the contribution by dissolved carbon dioxide was slight.

Chapter 2

The Theory of Buffer Action

2.1 Equilibrium aspects

An acid is a species such as the ammonium ion or the acetic acid molecule which has a tendency to lose a proton. A base (such as the ammonia molecule or the acetate anion) is able to accept a proton. Hence for every acid, HA, there is a conjugate base, A^-, and for every base, B, there is a conjugate acid, BH^+.

$$HA + H_2O \rightleftharpoons H_3O^+ + A^- \tag{2.1}$$

$$BH^+ + H_2O \rightleftharpoons H_3O^+ + B \tag{2.2}$$

Water is both a weak acid and a weak base,

$$H_2O + H_2O \rightleftharpoons H_3O^+ + OH^- \tag{2.3}$$

The strength of an acid, HA, in aqueous solution is expressed by its dissociation constant, K_a, where

$$K_a = (H^+)(A^-)/(HA) \tag{2.4}$$

the parentheses denoting activities of the hydrated species. Similarly for the conjugate acid of a base, B, K_a is given by

$$K_a = (H^+)(B)/BH^+) \tag{2.5}$$

It is often convenient to express the strengths of acids and bases in terms of pK_a values, where $pK_a = -\log K_a$. Extensive compilations of pK_a values of organic and inorganic acids and bases are available (Kortüm et al., 1961; Perrin, 1965, 1969, 1972).

Taking logarithms and rearranging, Equations 2.4 and 2.5 give

$$pH = pK_a + \log (\text{basic species})/(\text{acidic species}) \tag{2.6}$$

so that the pH of a solution containing equal activities of the species, A^- and HA, or B and BH^+, is equal to the pK_a value

4

of the acid or base. The pH of a solution containing a different ratio of conjugate acid and base can readily be obtained from Table 2.1. Using a 'practical' pK_a', as defined below (see Salt Effects), Equation 2.6 can also be written in the form

$$pH = pK_a' + \log \text{[basic species]} / \text{[acidic species]} \qquad (2.6a)$$

where the square brackets denote concentrations. This is the Henderson-Hasselbalch equation.

This equation can be regarded as the fundamental relation describing buffer equilibria. However, in assigning values to the concentration terms in this equation it is common to neglect the interaction of the acidic and basic species with the solvent water. This can lead to significant error outside the pH range 3 to 11, and it is tacitly assumed in the remainder of this Chapter that the equation will be used only within these limits. (For a fuller derivation and discussion of the Henderson-Hasselbalch equation, and its limitations, see Appendix IV.)

Quite generally, for the conjugate acid-base pairs involved in the equilibria,

$$H_n A^{x-} \rightleftharpoons H_{n-1} A^{(x+1)-} + H^+$$

and

$$H_{n+1} B^{(x+1)+} \rightleftharpoons H_n B^{x+} + H^+$$

$$pH = pK_a + \log(H_{n-1} A^{(x+1)-})/(H_n A^{x-}) \qquad (2.7)$$

$$pH = pK_a + \log(H_n B^{x+})/(H_{n+1} B^{(x+1)+}) \qquad (2.8)$$

Thus, from Equations 2.6, 2.7 and 2.8, the pH of a solution containing a mixture of a weak acid or base and its salt depends on the pK_a and the concentration ratio of the acidic and basic species. Because of the equilibria shown in Equations 2.1, 2.2 and 2.3, addition of small amounts of strong acid or alkali to these solutions is equivalent to the removal of small amounts of the weakly basic or acidic species, so that there is very little change in the pH of the solution. Such solutions are described as *buffers* and their effectiveness is expressed quantitatively by their *buffer capacity* (see below).

Many organic, and some inorganic, acids and bases have pK_a values between 2 and 12 so that, in principle, by partially neutralizing their solutions they could be used as buffers. Neglecting, for the moment, the effect of ionic strength, solutions of such an acid and its conjugate base in concentration ratios of 1 : 10 to 10 : 1 would furnish a series of buffers covering a pH range of $pK_a \pm 1$.

A solution of a weak acid, or its salt with a strong base, *alone*, is a poor buffer. This is also true of a weak base, or its salt with a mineral acid. In all these cases, the concentration ratio of acid and conjugate base differs markedly from unity so that the addition of strong acid or alkali leads to a rapid initial change in pH. However, the salt of a weak acid and a weak base, for example ammonium acetate, acts as a buffer, alone, because hydrolysis results in measurable amounts of the free acid and the free base in the solution.

2.2 Activity effects

The ionic strength, I, of a solution is given by the summation

$$I = \tfrac{1}{2} \sum (c_i z^2) \qquad (2.9)$$

where c_i is the concentration of each type of ion (in moles l^{-1}) and z is its charge.* Thus, for 0.15 M NaCl, $I = \tfrac{1}{2}(0.15 \times 1^2 + 0.15 \times 1^2) = 0.15$, and for 0.1M K_2SO_4, $I = \tfrac{1}{2}(0.2 \times 1^2 + 0.1 \times 2^2) = 0.3$. For solutions outside the pH range 4–10 the contributions of hydrogen and hydroxyl ions must also be included.

The activity of an ion, a_i, is related to its concentration,

$$a_i = c_i f_i \qquad (2.10)$$

---

*The *molar* scale, moles litre^{-1}, is designated by M to distinguish it from the *molal* scale, m, which is the number of moles dissolved in 1 kg of solvent. The two scales are very similar for dilute aqueous solutions, but are quite different for solutions in mixed solvents.

In dilute solutions, the activity coefficient, f_i, is given by the Debye-Hückel equation, an approximate form of which is (Davies, 1938)

$$-\log f_i = Az^2 I^{1/2}/(1 + I^{1/2}) - 0.1z^2 I \qquad (2.11)$$

where A is a constant which depends on the temperature, being 0.507 at 20°C, 0.512 at 25°C and 0.524 at 38°C.

This equation enables the pH of a buffer solution to be calculated from the thermodynamic pK_a of the buffer acid or base, and the concentrations of the buffer species:

$$pH = pK_a + \log [H_{n-1}A^{(x+1)-}]/[H_n A^{x-}] - (2x + 1)AI^{1/2}/(1 + I^{1/2})$$
$$+ 0.1(2x + 1)I \qquad (2.12)$$

$$pH = pK_a + \log [H_n B^{x+}]/[H_{n+1}B^{(x+1)+}] + (2x + 1)AI^{1/2}/(1 + I^{1/2})$$
$$- 0.1(2x + 1)I \qquad (2.13)$$

Example. What is the ionic strength of a pH 7.2 buffer comprising 36 ml 0.2M $Na_2 HPO_4$ and 14 ml 0.2M $NaH_2 PO_4$ in a final volume of 100 ml?

The concentration of HPO_4^{2-} is 0.072M
The concentration of $H_2 PO_4^-$ is 0.028M
The concentration of Na^+ is 2 x 0.072 + 0.028 = 0.172M

$$I = \tfrac{1}{2}(2^2 [HPO_4^{2-}] + 1^2 [H_2 PO_4^-] + 1^2 [Na^+])$$
$$= \tfrac{1}{2}(0.288 + 0.028 + 0.172) = 0.244$$

2.3 Effect of dilution

From the relations 2.12 and 2.13, the pH value of a buffer will change with its dilution, because of changes in the ionic strength. Table 2.2 shows the magnitude of the effect of diluting an equimolar solution of a HA/A^- buffer (total molar concentration stated) with an equal volume of water. The quantity $\Delta pH_{1/2}$ is defined as the increase in pH of a solution when it is diluted in this way. Dilution of acidic buffers increases the pH; with bases there is a decrease. Conversely, the addition of an 'inert' salt such as NaCl decreases the pH of acidic buffers and increases the pH of basic buffers.

Greater effects are produced when higher ionic charges are involved. For example, for a HA^-/A^{2-} buffer (such as KH_2PO_4, Na_2HPO_4) dilution with an equal volume of water increases the pH by approximately three times the amount shown in Table 2.2.

Because of its dependence on the squares of the charges of the ions present, the ionic strength of a buffer can vary over a wide range if di- or higher-valent ions are involved. Thus in the classical McIlvaine phosphate-citrate buffers the ionic strength varies 30-fold between pH 2.2 and pH 8.0. The difference in charge of an acid and its conjugate base causes the ionic strength of a buffer to vary with pH unless compensating amounts of a suitable salt are added. Hence, in calculating buffer tables it is necessary to use successive approximations for ionic strengths, activity coefficients and concentrations of the buffer species: computer-based iteration is convenient.

2.4 Salt effects

For buffers maintained at constant ionic strength by the addition of an 'indifferent' electrolyte, the ionic strength terms in Equations 2.12 and 2.13 are also constant and may be included with the pK_a term to give a 'practical' or 'conditional' pK_a' (equal to the pH of a solution containing equal *concentrations* of the two buffer species). Albert and Serjeant (1971) refer to these constants as 'mixed pK values', pK_a^M.

For a monoacidic base such as tris(hydroxymethyl)aminomethane (Tris), the 'practical' constant increases with ionic strength as shown in Table 2.3. For monobasic acids, such as acetic acid, a similar relation applies, except that pK_a' *decreases* by the same amounts. For aminoacids and similar zwitterionic species (see below) the pK_a' value above 7 decreases as ionic strength increases. (This is because the equilibrium is between a neutral species and an anion.)

Specific salt effects, in which one or more of the buffer species interacts with other components of the system, include precipitation and metal complexation. Examples are discussed in Chapter 4.

Valensi (1972) has proposed that in working with buffers of high ionic strength, pH measurements would lead to a better set of pH standards if liquid junction effects were largely eliminated by ensuring that a saturated KCl bridge was used with a pH standard solution having the same conductivity as the unknown solution. This applies if $3 < \text{pH} < 11$. Tables are given for KCl-rich solutions of phthalate, phosphate, borate, bicarbonate and hydroxide buffers from $0°C$ to $40°C$.

Examples (i) What is the effect on the pH of a buffer 0.005M in Tris and 0.005M in Tris hydrochloride if it is brought to an ionic strength of 0.1 by adding NaCl?

From Appendix III, the pK_a of Tris is 8.06 at $25°C$. The initial ionic strength of the Tris buffer is 0.005 and, from Table 2.3, $pK_a' = pK_a + 0.03 = 8.09$. Since the concentration ratio of Tris : Tris-HCl is unity, $pH = pK_a' = 8.09$. Changing I to 0.1 would change the pH to $pK_a + 0.11 = 8.17$, so the net result would be an increase of 0.08 in pH.

(ii) What is the pH of a solution containing 0.01M acetic acid and 0.02M sodium acetate at $25°C$?

The ionic strength of the solution is 0.02 and the pK_a of acetic acid is 4.76 (from Appendix III). Hence, from Table 2.3, its $pK_a' = 4.76 - 0.06 = 4.70$.

The concentration ratio of acetate ion to free acetic acid is $2 : 1$. Therefore, from Equation 2.6a,

$$pH = pK_a' + \log 2 = 4.70 + 0.30 = 5.00$$

(iii) If 1 ml of 1M HCl is now added to 100 ml of this buffer solution, what does the pH of the solution become?

Addition of hydrogen ions to a solution containing acetate ions leads to almost quantitative conversion to free acetic acid, because of the equilibrium, $H^+ + OAc^- \rightleftharpoons HOAc$. Allowing for dilution, free acetic acid concentration becomes

$0.02 \times 100/101\text{M}$ and the remaining acetate ion is $0.01 \times 100/101\text{M}$.

The ionic strength is now 0.01, so that pK_a' becomes $4.76 - 0.05 = 4.71$, and the $pH = 4.71 - \log(0.02/0.01) = 4.41$.

2.5 Ampholytes and Zwitterions

A substance such as m-aminophenol that possesses both a weakly acidic group and a weakly basic group is an amphoteric electrolyte or ampholyte. In this example, where the pK_a of the acidic group (here 9.9) is appreciably greater than the pK_a of the basic group (here 4.2) the neutral species is uncharged. On the other hand when, as with glycine, the pK_a of the acidic group (here 2.36) is appreciably smaller than the pK_a of the basic group (here 9.91), the neutral species carries both a positive and a negative charge. Such a species is known as a zwitterion. It is clear that $^+NH_3CH_2COO^-$ is ionic in character, so there is a problem as to how the activity effects of such zwitterions are to be dealt with. From freezing point measurements, the activity coefficient of glycine (on the molal scale) is

$$-\log \gamma = 0.0991 \, m - 0.0158 \, m^2 \qquad (2.14)$$

(Scatchard and Prentiss, 1934), so that at concentrations used in preparing buffers the log activity coefficient correction is small and approximates to one tenth of the molal concentration.

2.6 Buffer capacity

The pH of a buffer system based on the equilibrium,

$HA \rightleftharpoons H^+ + A^-$, is given by:

$$pH = pK_a' + \log [A^-]/[HA] \qquad (2.15)$$

Given the relations that the sum of $[A^-] + [HA]$ remains constant at some concentration, c, and that the sum of the cation concentrations ($[H^+] + [K^+]$, etc) is equal to that of the anions ($[OH^-] + [A^-]$), it can readily be deduced that the increment of strong base needed to change the pH of the buffer by a small amount is given by:

$$\frac{d[B]}{dpH} = 2.303 \left\{ \frac{K_a' c [H^+]}{(K_a' + [H^+])^2} + [H^+] + \frac{K_w}{[H^+]} \right\} \qquad (2.16)$$

where K_w is the ionic product of water. The second and third terms, which are important in strongly acid and alkaline solutions, respectively, give the buffering ability of the solvent, water. In the region between pH 3 and pH 11 these terms are negligible and the maximum value of $d[B]/dpH$ is

$$\beta_{max} = 2.303\, c/4 = 0.576c$$

found at pH = pK_a', that is, when $[A^-]$ and $[HA]$ are equal. This corresponds to the point of half-neutralization in titrations of weak acids or bases. At this point the slope of the titration curve is a minimum.

The quantity, $d[B]/dpH$, is a quantitative measure of the buffering ability of a solution. It was introduced by Van Slyke (1922) as the 'buffer unit' or 'buffer value', β, and is also known as the 'buffer capacity'. It is the reciprocal of the slope of the pH-neutralisation curve.

Example Compare the buffer capacities at pH 7.4 of the Tris buffer given in Table 10.32 with the HEPES buffer given in Table 3.6.

Equation 2.16 applies (see below). For Tris, at pH 7.4, $c = 0.05$M, [Tris-HCl] = 0.042M, hence $I = 0.042$ and $pK_a' = 8.14$.

Hence $K_a' = 7.25 \times 10^{-9}$, $[H^+] = 3.98 \times 10^{-8}$. Substitution into Equation 2.16 gives $d[B]/dpH = 0.015$.

For HEPES, $pK_a' = 7.55$ and $c = 0.05$M. A similar calculation gives $d[B]/dpH = 0.028$. Hence HEPES is a more effective buffer than Tris at this pH.

2.6.1 Buffer capacity of a polybasic acid.

For a dibasic acid, β is given by

$$\beta = 2.303 \left\{ \frac{c[\mathrm{H^+}]\left(\dfrac{1}{K_1} + \dfrac{4[\mathrm{H^+}]}{K_1 K_2}\right)}{\left(1 + \dfrac{[\mathrm{H^+}]}{K_1} + \dfrac{[\mathrm{H^+}]^2}{K_1 K_2}\right)^2} + [\mathrm{H^+}] + \frac{K_w}{[\mathrm{H^+}]} \right\} \tag{2.17}$$

where $pK_1 > pK_2$, and K_1, K_2 are practical constants.

For tribasic acids, such as citric acid, the equation is

$$\beta = 2.303$$

$$\frac{c[\mathrm{H^+}]\left(\dfrac{1}{K_1} + \dfrac{4[\mathrm{H^+}]}{K_1 K_2} + \dfrac{9[\mathrm{H^+}]^2}{K_1 K_2 K_3} + \dfrac{[\mathrm{H^+}]^2}{K_1{}^2 K_2} + \dfrac{4[\mathrm{H^+}]^3}{K_1{}^2 K_2 K_3} + \dfrac{[\mathrm{H^+}]^4}{K_1{}^2 K_2 K_3}\right) + [\mathrm{H^+}] + \dfrac{K_w}{[\mathrm{H^+}]}}{\left(1 + \dfrac{[\mathrm{H^+}]}{K_1} + \dfrac{[\mathrm{H^+}]^2}{K_1 K_2} + \dfrac{[\mathrm{H^+}]^3}{K_1 K_2 K_3}\right)^2} \tag{2.18}$$

The same equations apply to weak bases and their conjugate acids. Values of β are additive for mixtures of buffers.

The shapes of some typical buffer capacity curves with varying pH are shown in Figures 2.1 and 2.4. For a monobasic acid type of buffer the variation of β with pH is given in Table 2.4 over the range $pK_a \pm 2$. It is apparent from the Table that outside the range $pK_a \pm 1$ there is little buffer ability and that if 50% or more of the maximum buffering capacity is to be realized the pH range is limited to $pK_a \pm 0.75$.

The pH of a solution of a 1 : 1 salt of a weak acid and a weak base is given by

$$pH = \tfrac{1}{2}(pK_1 + pK_2) \tag{2.19}$$

where the pK_a values are for the acidic and basic groups. The buffer capacity is

$$\beta = 2.303\, c\, [\mathrm{H^+}] \left\{ \frac{K_1}{(K_1 + [\mathrm{H^+}])^2} + \frac{K_2}{(K_2 + [\mathrm{H^+}])^2} \right\} \tag{2.20}$$

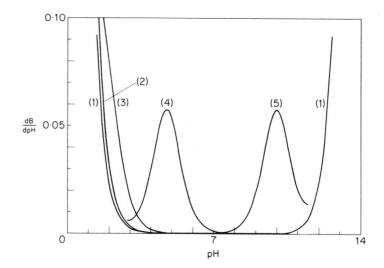

Fig. 2.1 Buffer capacities of aqueous solution

(1) water alone, (2) 0.1M CCl_3COOH (pK$_a$ 0.66), (3) 0.1 M H_2SO_4 (pK$_a$ 1.86), (4) 0.1M HOAc (pK$_a$ 4.76), (5) 0.1M phenol (pK$_a$ 10.00).

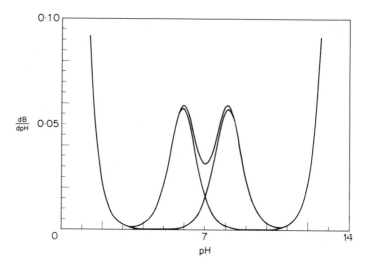

Fig. 2.2 Cumulative effects of buffer capacities.

Curves for 0.1M sodium hydrogen maleate (pK 5.94) and Tris (pK 8.14) singly and in admixture.

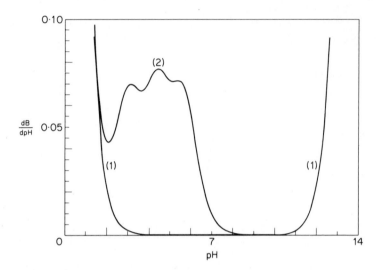

Fig. 2.3 Buffer capacity curves for 0.1M citric acid and water.

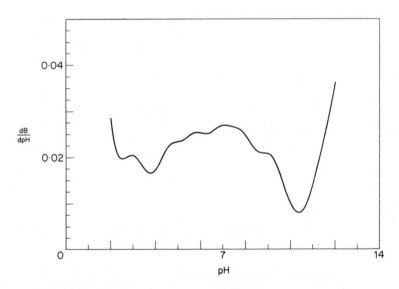

Fig. 2.4 Buffer capacity curve for Britton-Robinson universal buffer stock solution.

The greatest value of β for such a salt is reached when pK_1 and pK_2 are equal. Values for a 0.1M solution are 0.115, 0.038, 0.0045 and 0.0047 for differences between pK_1 and pK_2 of 0, 2, 4 and 6 pH units respectively.

2.7 Pseudo buffers

Strongly acidic or basic solutions show little change in pH when acid or alkali is added. This is a consequence of the nature of the pH scale and of the ionization of water as a weak base ($pK_a \sim 0$) and as a weak acid ($pK_a \sim 14$). Such solutions, in which the buffer action is due to the solvent rather than to any added solute, are not ordinarily described as 'buffers' but it is convenient to speak of them as 'pseudo buffers'.

The buffer capacity of water containing completely dissociated acids or alkalis is

$$d\,[B]/dpH = 2.303([H^+] + K_w/[H^+])\qquad(2.21)$$

Because of the high concentrations of solvent molecules, this buffer capacity may be much larger than for conventional buffers.

2.8 Self buffers

Sodium tetraborate, $Na_2B_4O_7$, dissociates to form an equimolar concentration of metaboric acid, HBO_2, and metaborate ion, BO_2^-, establishing such a reproducible hydrogen ion concentration in the solution that this material is used as a primary standard in pH measurement. (This interpretation is oversimplified: account should also be taken of the tendency of boric acid and borate ion to form concentration-dependent polymers.) Another example is potassium tetroxalate which is an equimolar mixture of potassium hydrogen oxalate and oxalic acid.

It is convenient to designate as 'self buffers' substances which possess this ability to act as buffers. They are very convenient as pH standards, particularly as the solutions are rather insensitive to variations in concentration if the pH lies between about 3 and 11.

A much wider range of self buffers is encompassed when partially neutralized salts of substances with two or more pK_a values are involved. Phthalic acid is dibasic, with pK_a values at 25°C of 2.95 and 5.41; potassium hydrogen phthalate solutions have a pH roughly midway between these values (slightly lower because of ionic strength effects) and the buffer capacity is as expected for a solution 1.2 pH units removed from the pK_a. This is the basis of the use of 0.05M potassium hydrogen phthalate as the reference standard in pH measurement. Another example is potassium hydrogen tartrate (pK_a values of parent acid, 3.04 and 4.37) which has a better buffer capacity but is not very soluble. Sodium hydrogen malate (pK_a values of parent acid, 3.40 and 5.13) and potassium dihydrogen citrate (pK_a values of 3.13 and 4.76) could also be useful. The pH of a 0.05m KH_2 citrate solution is 3.68 at 25°C. A 0.2M solution of sodium hydrogen diglycollate (pK_a values 2.96 and 4.43) has a pH = 3.40 ± 0.02 over the temperature range 10–35°C, has a slight $\Delta pH_{1/2}$ and has been proposed as a pH reference buffer (Keyworth and Hahn, 1958). On the other hand, the pK_a separations are too great for a salt such as potassium dihydrogen phosphate (pK_a values of phosphoric acid are 2.16, 7.21, 12.33) to be useful as a self buffer.

Mention has already been made of the self buffering by a salt of a weak acid and a weak base, such as ammonium acetate (pK_a values of 4.76 and 9.25). In this example the pH of a solution is near to 7 ($\approx \frac{1}{2}(pK_{HOAc} + pK_{NH_3})$), but the buffer capacity is small because the pH is more than 2 units away from either pK_a value. (In analytical chemistry this disadvantage is partially overcome by using high concentrations). Ammonium bicarbonate (pK_a values of 6.35, 9.25) or diammonium hydrogen phosphate (pK_a values 7.20, 9.25) giving pH values of approximately 7.4 and 8.0, respectively, for 0.05M solutions, have much better buffer capacities.

Piperazine phosphate monohydrate, $C_4H_{12}N_2HPO_4 \cdot H_2O$, is a very good self buffer (pK_a values of piperazine 5.333, 9.731; relevant pK_a value of phosphoric acid 7.198 at 25° C) and it has been proposed for use as a pH standard. Measured pH values for an 0.02m solution from 0 to 50° C are given in Table 2.5. For an 0.05m solution the pH values should be increased by 0.009.

Ampholytes (zwitterions) include an extensive range of substances that could be used as self buffers. Table 2.6 gives a list of isoionic ampholytes proposed for use as buffers in protein fractionation in a natural pH gradient (Svensson, 1962). Solutions of these substances in water have pH values close to the listed value of pI and when pI - pK_1 is less than 1.5 they can be considered to be self buffers.

2.9 Mixtures of buffers

The effective buffer range for a weak acid or base is approximately from $pH = pK_a + 1$ to $pH = pK_a - 1$. When two or more buffers are present, the effects are additive so that the buffering ability is spread over a wider pH range. Examples are McIlvaine's (1921) citric acid-phosphate mixtures for pH 2.6–8 (Table 10.45) and Smith and Smith's (1949) piperazine-glycylglycine mixtures for pH 4.4–10.8 (Table 10.46). The pK_a of ethanolamine (9.5) falls conveniently between two of the pK_a values of phosphoric acid (7.2, 12.3) so that ethanolamine-phosphate mixtures provide almost uniform buffer capacity between pH 6.7 and pH 12.8 (Thies and Kallinich, 1953).

If a buffer system has several successive pK_a values which differ by about 2 pH units, approximately linear buffer capacity results. This property has been exploited in 'universal' buffers having high buffer capacities over a wide pH range. Britton and Robinson (1931) used equimolar mixtures based on seven pK_a values of citric, phosphoric, boric and diethylbarbituric acids to cover the pH range 2.6–12, as listed in Tables 10.47 and 10.48. Coch Frugoni (1957) gave tables adapting these buffers to constant ionic strengths 0.005 and 0.02 by varying the amount of water added, and to $I = 0.1$, 0.5 and 1.0 by addition of sodium chloride.

2.10 Temperature dependence

The effect of temperature on the pH of a buffer solution depends on the temperature dependence of the activity coefficient terms and of the pK_a of the buffer species. The latter is usually much the more important. For systems of the type, B/BH^+, the effect of temperature on the pK_a is given, to a first approximation, by

$$\frac{-d(pK_a)}{dT} = \frac{pK_a - 0.9}{T} \tag{2.22}$$

where T is in $^\circ K$ (Perrin, 1964). For dications, the equation is

$$\frac{-d(pK_a)}{dT} = \frac{pK_a}{T} \tag{2.23}$$

Around $25^\circ C$, for a base such as piperidine ($pK_a = 11.12$), the pK_a decreases by 0.034 units per degree. For carboxylic acids around ambient temperatures, on the other hand, changes in pK_a values are much smaller. The temperature coefficients for phenols are also smaller than for bases having comparable pK_a values.

2.11 Effect of pressure on buffers

High pressure increases the ionization of weak electrolytes, by enhancing the solvation of the ions, but the effect is not very great. At $25^\circ C$ and 3000 atmospheres, pK_a values are decreased as follows: formic acid, by 0.38, acetic acid, by 0.50, and propionic acid, by 0.55 (Hamann and Strauss, 1955). The logarithm of the basic dissociation constant of ammonia (into ammonium ion and hydroxyl ion) is increased by 1.14 for a pressure increase of 3000 atmospheres (Buchanan and Hamann, 1953) and by 2.72 for a pressure increase of 12 000 atmospheres (Hamann and Strauss, 1955). Increases for organic bases at 3000 atmospheres are 1.05 for methylamine, 0.67 for dimethylamine, and 0.71 for tri-methylamine (Hamann and Strauss, 1955). The dissociation constant of water, pK_w, decreases by 0.51 at 25°C if the pressure is raised to 2000 atmospheres (Hamann, 1963.)

2.12 Further reading

A more detailed discussion of many of the topics discussed in this chapter is given by Bates (1964).

Table 2.1 *Calculation of concentration ratios from $pH = pK_a' + log$ [basic form]/[acidic form]*

$pH = pK_a'$ plus:	$\dfrac{[Base]}{[Acid]}$	$\dfrac{[Base]}{[Acid] + [Base]}$	$pH = pK_a'$ minus:	$\dfrac{[Base]}{[Acid]}$	$\dfrac{[Base]}{[Acid] + [Base]}$
1.00	10.00	0.91	0.10	0.79	0.44
0.90	7.95	0.89	0.20	0.63	0.39
0.80	6.31	0.86	0.30	0.50	0.33
0.70	5.01	0.83	0.40	0.40	0.28
0.60	3.98	0.80	0.50	0.32	0.24
0.50	3.16	0.76	0.60	0.25	0.20
0.40	2.51	0.72	0.70	0.20	0.17
0.30	2.00	0.67	0.80	0.16	0.14
0.20	1.59	0.61	0.90	0.13	0.11
0.10	1.26	0.56	1.00	0.10	0.09
0.00	1.00	0.50			

Table 2.2 *Dilution values ($\Delta pH_{1/2}$) for equimolar acid buffer solutions (calculated from Davies' equation)*

Molarity	$\Delta pH_{1/2}$	Molarity	$\Delta pH_{1/2}$	Molarity	$\Delta pH_{1/2}$
0.100	0.024	0.050	0.021	0.025	0.017
0.020	0.016	0.010	0.012	0.005	0.009

Table 2.3 *Effect of ionic strength on pK_a' values of acids and bases**

$pK_a' = pK_a +$ correction (for bases)

$\qquad = pK_a -$ correction (for acids and zwitterionic bases)

For a system $BH^{n+} \rightleftharpoons B^{(n-1)+} + H^+$, or for $HA^{(n-1)} \rightleftharpoons A^{n-} + H^+$, the term in column 3 must be multiplied by $(2n - 1)$.

I	$I^{1/2}$	Correction	I	$I^{1/2}$	Correction
0.001	0.032	0.015	0.035	0.187	0.076
0.002	0.045	0.021	0.040	0.200	0.080
0.003	0.055	0.026	0.045	0.212	0.084
0.004	0.063	0.030	0.050	0.224	0.087
0.005	0.071	0.033	0.060	0.245	0.093
0.006	0.077	0.036	0.070	0.265	0.099
0.007	0.084	0.038	0.080	0.283	0.103
0.008	0.089	0.041	0.090	0.300	0.108
0.009	0.095	0.043	0.100	0.316	0.111
0.010	0.100	0.045	0.200	0.447	0.136
0.015	0.122	0.054	0.300	0.548	0.149
0.020	0.141	0.061	0.400	0.632	0.156
0.025	0.158	0.066	0.500	0.707	0.159
0.030	0.173	0.072			

*Calculated using Davies' equation.

Table 2.4 *Dependence of buffer capacity, β, of HA/A^- buffer on pH*

pH = pK_a minus:	$\%\beta_{max}$	pH = pK_a plus:	$\%\beta_{max}$
2.0	3.9	0.1	98.6
1.9	4.9	0.2	94.8
1.8	6.1	0.3	88.9
1.7	7.7	0.4	81.4
1.6	9.6	0.5	73.0
1.5	11.9	0.6	64.1
1.4	14.7	0.7	55.4
1.3	18.2	0.8	47.2
1.2	22.3	0.9	39.7
1.1	27.2	1.0	33.0
1.0	33.0	1.1	27.2
0.9	39.7	1.2	22.3
0.8	47.2	1.3	18.2
0.7	55.4	1.4	14.7
0.6	64.1	1.5	11.9
0.5	73.0	1.6	9.6
0.4	81.4	1.7	7.7
0.3	88.9	1.8	6.1
0.2	94.8	1.9	4.9
0.1	98.6	2.0	3.9
0.0	100.0		

Table 2.5 *pH Values of 0.02 molal piperazine phosphate solutions**

T, °C	pH	T, °C	pH
0	6.582	30	6.235
5	6.514	35	6.186
10	6.451	40	6.141
15	6.394	45	6.098
20	6.337	50	6.056
25	6.285		

*Hetzer, *et al.* (1968)

Table 2.6 *Isoionic pH values of ampholytes**

Zwitterionic species	pI† (25°C)
Aspartic acid	2.77
Glutathione	2.82
Aspartyl-tyrosine	2.85
o-Aminophenylarsonic acid	3.00
Aspartyl-aspartic acid	3.04
p-Aminophenylarsonic acid	3.15
Picolinic acid‡	3.16
L-Glutamic acid	3.22
β-Hydroxyglutamic acid	3.29
Aspartyl-glycine	3.31
Isonicotinic acid‡	3.35
Nicotinic acid	3.44
Anthranilic acid	3.51
p-Aminobenzoic acid	3.62
Glycyl-aspartic acid	3.63
m-Aminobenzoic acid	3.93
Diiodotyrosine‡	4.29
Cystinyl-diglycine‡	4.74
α-Hydroxyasparagine‡	4.74
α-Aspartyl-histidine‡	4.92
β-Aspartyl-histidine‡	4.94
Cysteinyl-cysteine‡	4.96
Pentaglycine‡	5.32
Tetraglycine‡	5.40
Triglycine‡	5.59
Tyrosyl-tyrosine‡	5.60
Isoglutamine‡	5.85
Lysyl-glutamic acid‡	6.10
Histidyl-glycine	6.81
Histidyl-histidine	7.30
Histidine	7.47
L-Methylhistidine	7.67
Carnosine	8.17
α, β-Diaminopropionic acid	8.20
Anserine	8.27
Tyrosyl-arginine	8.53
L-Ornithine	9.70
Lysine	9.74
Lysyl-lysine	10.04
Arginine‡	10.76

*Svensson (1962)
†pI is the pH of the isoionic species and is approximately equal to $\frac{1}{2}(pK_1 + pK_2)$.
(Range of pI − pK_1 values is 0.34 to 2.43)
‡Values of pI − pK_1 exceed 1.5.

Table 2.7 *Components of universal buffers*

Boric acid, phosphoric acid, phenylacetic acid.[*]
Acetic acid, phenolsulphonic acid, phosphoric acid.[†]
Boric acid, citric acid, diethylbarbituric acid, phosphoric acid.[‡]
Boric acid, phosphoric acid, phthalic acid.[§]
Boric acid, citric acid, phosphoric acid.[‖]
Carbonic acid, citric acid, phosphoric acid, 2-amino-2-methyl-propane-1,3-diol.[¶]

[*]Prideaux and Ward (1924)
[†]Mellon *et al* (1921)
[‡]Britton and Robinson (1931)
[§]Best and Samuel (1936)
[‖]Teorell and Stenhagen (1938); Carmody (1961)
[¶]Ellis (1961)

Chapter Three

Applications of pH Buffers

3.1 Factors governing the choice of a buffer

The buffering ability of a weakly acidic or basic group is limited approximately to the range, $pH = pK_a \pm 1$, the greatest effect being at $pH = pK_a$. This is clearly the most important single factor in choosing a buffer for any particular application and reference to lists of pK_a values such as those given in Appendix III provides a rapid indication of possible buffer substances.

However, there are usually a number of other considerations such as the nature of the reaction system, the ionic strength, the effects of temperature change and dilution, and the possibility of forming insoluble or strongly coloured species or otherwise interacting with components of the solution. For example, the effect of temperature on the pK_a of the buffer should be known and, preferably, should not be very large. As pointed out in Chapter 4, this is important for buffers based on Tris (tris(hydroxymethyl)aminomethane) and aliphatic amines. The buffer species should also be chemically stable, should be readily soluble in water and not be readily extracted by organic solvents.

A good buffer shows little change in pH if there is accidental contamination with acidic or basic materials. For a buffer of the type, $BH^+ \rightleftharpoons B + H^+$, acidic contaminants have little effect if they are weaker acids than BH^+. Similarly, basic contaminants that are weaker bases than B are not important. For measurements of rates or of physical constants it is often desirable to work under conditions of known low ionic strength. In all cases, the pH of a buffered medium should be determined only after all components have been added and at the temperature of the final measurements. Special requirements for particular buffer applications are discussed below under separate headings.

3.2 Measurement of pH

A pH determination using the hydrogen gas electrode requires two measurements of the e.m.f. of a pH cell, E_{std} and E_x, corresponding to the immersion of the electrodes in a solution of known pH and in the unknown solution. The relation is

$$pH_{(x)} = pH_{(std)} + (E_x - E_{std})F/(2.303RT) \qquad (3.1)$$

However, almost all pH measurements now use the glass electrode and, to allow for slight variations in its pH response, the electrode is calibrated against two pH standards which, if possible, bracket the pH range of the test solutions. The measured e.m.f. of an unknown solution can then be converted directly to a pH measurement. This procedure is based on the operational definition of pH as proposed by Bates and Guggenheim (1960). The equation is

$$pH_{(x)} = pH_{(std\ 1)} + (pH_{(std\ 2)} - pH_{(std\ 1)})\frac{(E_x - E_{std\ 1})}{(E_{std\ 2} - E_{std\ 1})} \qquad (3.2)$$

Alternatively, the pH of an unknown solution may be determined with the aid of indicators whose colour changes have been correlated with pH values of standard solutions. For fuller details of the technique of pH measurement, see Albert and Serjeant (1971).

Determinations of pH, either by potentiometry or by spectrophotometry, thus require accurately prepared reference solutions of analytical grade reagents as standards. The standards should be well separated on the pH scale but, to minimize the effects of liquid-junction potentials, they should not be less than pH 3 nor greater than pH 11.

The usual standards for pH measurement are 0.05M potassium hydrogen phthalate and 0.01M sodium borate buffers having the assigned values given in Table 3.1. (Strictly, the United States National Bureau of Standards (NBS) standards are based on molalities (moles per kg.

solvent) and the British standards are on the molarity scale (moles per litre of solution), but the differences are negligible for such dilute aqueous solutions: the two phthalate buffer scales agree to within an accuracy of ±0.005 pH units.) Alternative pH standards are given in Table 3.2.

These standards are a compromise between concentration and buffer capacity. If the concentration is too low the buffer capacity is poor, but if the concentration is too high the basis for defining the pH value becomes less certain because of greater difficulty in justifying the value assigned to the single ion activity coefficient of the chloride ion.

Buffer pH values on the original Sörensen psH scale can be converted approximately to the agreed standard scale by adding 0.04.

Outside the pH range 3–11, measurements can still be made with the usual glass electrode, reference electrode assembly, although their theoretical significance may be decreased because of uncertainties about liquid junction potentials and the cation response of the glass electrode in alkaline solutions. Convenient reference buffers for high pH measurements have been described (Tuddenham and Anderson, 1950), based on the insolubility of calcium hydroxide and calcium carbonate. If a small amount of acid or carbon dioxide is added to a calcium chloride solution containing a slurry of calcium hydroxide some of the dissolved calcium hydroxide is neutralized and more of the solid material dissolves. Addition of alkali precipitates some calcium hydroxide, so that approximate pH constancy is maintained. The reported pH value for 0.2M calcium chloride saturated with calcium hydroxide is 11.88 at 25°C and the temperature coefficient is −0.03 pH unit/°C. In a solution also 2M in sodium chloride the pH at 20°C was 11.97. Similar buffers should be possible with barium, magnesium and strontium chlorides and their hydroxides.

The Clark and Lubs (1916) series of buffers are convenient for colorimetric pH measurement between pH 1.0 and 10.2 using acid-base indicators. This series, with more recently determined pH values, comprises Tables 10.5, 10.10, 10.20, 10.25 and 10.37. An alternative set, for the pH range 7.0–13.0, is afforded by Tables 10.32, 10.37a, 10.41, 10.42, 10.43 and 10.44.

For pH measurements outside the usual range it may be convenient to use 0.05M potassium tetroxalate (pH = 1.672 at 15°C, 1.679 at 25°C, 1.688 at 35°C) or calcium hydroxide solution saturated at 25°C (pH = 12.810 at 15°C, 12.454 at 25°C, 12.133 at 35°C). Values up to 95°C for both of these standards are given by Bates (1962).

The pH 7.4 phosphate buffer given in Table 3.3 is useful as a reference in measuring the pH of blood. The following buffer (ionic strength 0.13) is also suitable: 0.01M KH_2PO_4 (1.360g/l) and 0.04M Na_2HPO_4 (5.677 g/l) in 1 litre of CO_2-free water. At 37.5°C and 38°C its pH is 7.416 ± 0.004 (Semple *et al.*, 1962). An alternative buffer is 1.816 g KH_2PO_4 and 9.501 g $Na_2HPO_4 \cdot 2H_2O$ in 1000 g of water: at 38°C its pH is 7.381 ± 0.002 (Spinner and Petersen, 1961). To avoid complex formation with ionic species such as calcium and magnesium ions in blood it may be preferable to use a Tris buffer. A mixture 0.01667m in Tris and 0.0500m in Tris-HCl has a pH of 7.699 at 25°C and 7.382 at 37°C, $(d(pH)/dT = -0.026)$ (Durst and Staples, 1972). Values are given in this reference for the range 0–50°C. If the solution is also made 0.11m in NaCl to give an isotonic solution ($I = 0.16$), the pH of the buffer is increased by 0.045 at 25°C and 37°C.

Measurement of pH* values in partially aqueous solutions and pD in heavy water, is described in Chapter 6.

3.3 Biochemistry and biology

The original concept of pH-buffer action arose out of biochemical studies, and the need for pH control in all aspects of biological research is now universally recognized. Unfortunately, until recently there were few suitable substances having good buffering capacity in the physiologically important pH range 6 to 8, so that a buffer substance had to be selected almost entirely on the basis of its pK_a value with little regard for its compatibility with the biological system under study. As discussed in Chapter 4 there are disadvantages associated with the commonly used phosphate, borate and bicarbonate/CO_2 buffers, including the complication that in many systems phosphate or bicarbonate is an essential metabolite that is consumed as reactions proceed.

In biological studies, either with isolated enzyme systems or with intact tissue, it is obviously desirable to confirm that added materials such as buffer components do not exert any spurious effects. It is important to carry out measurements in more than one kind of buffer in order to establish which buffers cause least disturbance to the system under study, the usual criterion being that the observed reaction rates or transformations are maximal. The recent extension of the number of buffers covering the physiological pH range now makes it possible to assess much more effectively the suitability of possible buffer systems.

Typical phosphate and borate buffers are listed in Chapter 10, while Table 3.4 gives the concentrations of $NaHCO_3$ at 37°C in equilibrium at various pH values with a gas phase containing 5% CO_2. The calculations are based on the Henderson-Hasselbalch equation. For temperatures between 20°C and 37°C the bicarbonate concentrations should be increased by 1.88% per degree. Because the reversible hydration of CO_2 is a rather slow process, it may be advantageous to include a trace of the enzyme carbonic anhydrase in systems where bicarbonate/CO_2 buffers are used, so as to maintain more nearly equilibrium conditions.

The importance of the bicarbonate/CO_2 equilibrium in biological studies has led to the development of 'CO_2 buffers' (typically for use in the centre wells of manometric vessels) capable of maintaining a virtually constant pressure of CO_2 in an enclosed gas phase. A suitable 'CO_2 buffer' is an aqueous solution of diethanolamine (containing dithizone or thiourea to inhibit auto-oxidation) in which the following equilibrium is set up:

$$HN(CH_2CH_2OH)_2 + CO_2 + H_2O \rightleftharpoons HCO_3^- + H_2N^+(CH_2CH_2OH)_2$$

For a discussion of the theory and use of such buffers, see Krebs (1951).

One of the earliest extensions of the list of possible buffers for biological work was Gomori's (1946) proposal of 2,4,6-trimethylpyridine, tris(hydroxymethyl)aminomethane and 2-amino-2-methyl-1,3-propanediol. The use of peptides such as glycylglycine and, more particularly, N-dimethyl-leucylglycine (Léonis, 1948) as biological buffers is now of little more than historical interest because a much wider range of suitable, readily water-soluble buffer substances having soluble calcium salts has been developed.

The choice of alternative buffers has increased with the commercial availability of zwitterionic amino acids, mainly N-substituted taurines or N-substituted glycines (Good *et al.*, 1966) so that organic buffers suitable for use in biochemistry now include:

ACES $NH_2COCH_2NHCH_2CH_2SO_3H$ (M.Wt. 182.2)

ADA $NH_2COCH_2N(CH_2COOH)_2$ (M.Wt. 190.2)

BES $(HOCH_2CH_2)_2NCH_2CH_2SO_3H$ (M.Wt. 213.3)

Bicine $(HOCH_2CH_2)_2NCH_2COOH$ (M.Wt. 163.2)

Bis-tris $(HOCH_2CH_2)_2NC(CH_2OH)_3$ (M.Wt. 209.3)

Cholamine $(CH_3)_3^+NCH_2CH_2NH_2Cl^-$ (M.Wt. 138.6)
   chloride

EPPS $HOCH_2CH_2N\!\!\!\bigcirc\!\!\!NCH_2CH_2CH_2SO_3H$ (M.Wt. 252.3)
(or HEPPS)

HEPES $HOCH_2CH_2N\!\!\!\bigcirc\!\!\!NCH_2CH_2SO_3H$ (M.Wt. 238.3)

MES $O\!\!\!\bigcirc\!\!\!NCH_2CH_2SO_3H$ (M.Wt. 195.2)

Mono-tris $HOCH_2CH_2NHC(CH_2OH)_3$ (M.Wt. 165.2)

MOPS $O\langle\quad\rangle NCH_2CH_2CH_2SO_3H$ (M.Wt. 209.3)

PIPES $HO_3SCH_2CH_2N\langle\quad\rangle NCH_2CH_2SO_3H$ (M.Wt. 302.4)

TAPS $(HOCH_2)_3CNHCH_2CH_2CH_2SO_3H$ (M.Wt. 243.2)

TES $(HOCH_2)_3CNHCH_2CH_2SO_3H$ (M.Wt. 229.2)

Tricine $(HOCH_2)_3CNHCH_2COOH$ (M.Wt. 179.2)

The pK_a' data listed in Table 3.5 enable the compositions of constant ionic strength buffer solutions to be calculated using Table 10.1. If the ionic strength is other than 0.1 the 'practical' pK_a' values can be corrected using Table 2.3. Alternatively, the computer programmes given in Chapter 5 allow buffer tables to be constructed for a wide range of conditions. For example, Table 5.3 has been used to calculate the buffer compositions between pH 6.6 and 8.5 for an 0.05M HEPES solution at 20°C which had been titrated with 1M NaOH and adjusted to $I = 0.1$ with 1M NaCl. Results are given in Table 3.6. These values also apply to other zwitterionic buffers if the pH values are corrected as set out in Table 3.7.

Several of these buffers (including MES, PIPES, Cholamine, BES, TES and HEPES) have the additional advantage that they form only weak complexes with Mg, Ca or Mn ions and hence are useful in kinetic studies of enzymes that require these ions. Information on metal-binding by other members of the series is given in Table 7.5.

Growth of mammalian cells in culture media is very sensitive to the pH of the media and also depends on the kinds of buffer species present. Also, a high buffer capacity but a low buffer concentration is needed if the pH is not to fluctuate undesirably as a result of tissue metabolism, leading to inhibition of growth and to undesirable effects on other cell parameters. Some of the earliest buffered media for tissue metabolism studies were Krebs-Ringer-phosphate and bicarbonate solutions (Krebs and Henseleit, 1932), comprising mixtures of inorganic salts buffered by a larger amount of sodium phosphate and sodium bicarbonate, respectively. Media for growing micro-organisms are similar but frequently also include citrate to prevent precipitation of calcium phosphate and, perhaps, to provide a carbon nutrient source. (For example, Davis's minimal medium (Davis and Mingioli, 1950)). Disadvantages and limitations of bicarbonate/CO_2 and phosphate buffers are well known, while Tris buffers are often cytotoxic. Of these zwitterionic buffers HEPES has been found the most generally useful, for example with Eagle's basal medium, for the culture of animal tissues and cells. Use of HEPES as a tissue culture buffer was also studied by Shipman (1969).

Improved pH control between pH 6.7 and 8.4 was obtained when 20–50mM HEPES and 5–10mM Tricine were incorporated into Eagle's (1959) culture medium (Ceccarini and Eagle, 1971). Subsequently (Eagle, 1971), smaller pH fluctuations were found when a mixture 10mM in BES (or MOPS), 15mM in HEPES and 10mM in EPPS was added to the culture medium to supplement the 24mM sodium bicarbonate originally used as a buffer. Recommended buffer combinations for media at specified pH values are also given in this paper. However, as buffer species vary in their toxicity towards different types of cells, care must be taken in choosing buffers to use with new lines of cells. Preliminary studies should be carried out with several possible buffers to establish which, if any, are suitable. The possibility of using polyelectrolytes such as polymethacrylic acid as pH and pCa buffers in cell culture studies has recently been examined. (Matsumura *et al.*, 1968). Amberlite IRC 50 gave strong buffering of these ions under physiological conditions.

In biological systems where effects of K^+ and Na^+ may be important, buffers lacking these ions may be desirable; such buffers may comprise amines or zwitterions in combination with organic acids.

Example What strength of a buffer of known pK_a' is needed if the initial pH of a solution is to change by less than a specified amount if some acidic material is added or generated?

If the initial concentrations of acidic and basic buffer species are C_a M and C_b M, and the initial pH is pH_1, then

$$pH_1 = pK_a' + \log(C_b/C_a)$$

so that

$$C_b/C_a = 10^{(pH_1 - pK_a')} = \alpha \qquad (3.3)$$

If an amount of acid giving a concentration x M in the solution is added, C_b changes to $C_b - x$ and C_a increases to $C_a + x$. The solution is now at pH_2, and

$$pH_2 = pK_a' + \log((C_b - x)/(C_a + x))$$

so that

$$(C_b - x)/(C_a + x) = 10^{(pH_2 - pK_a')} = \beta \qquad (3.4)$$

Rearranging (3.3) and (3.4) and solving, gives

$$C_a = (\beta + 1)x/(\alpha - \beta)$$

$$C_b = \alpha(\beta + 1)x/(\alpha - \beta)$$

For example, a HEPES ($pK_a' = 7.55$) buffer of pH 7.8 at $20°C$, $I = 0.1$, would need to have $C_a = 0.016M$ and $C_b = 0.029M$ if the addition of an equal volume of $0.01M$ acetic acid was to lower the pH by less than 0.4.

3.4 Spectroscopy

Sets of buffers that are nearly transparent down to 240 nm have been developed for use in ultraviolet and visible spectroscopy (Perrin, 1963). Their low and constant ionic strength of 0.01 and pH range (2.2 to 11.6) make them suitable for spectrophotometric pK_a determinations because only small, constant corrections are needed to convert experimental pK_a values to thermodynamic ones. Buffer compositions are given in Table 3.8. Other buffers suitable for ultraviolet spectroscopy include those based on *N*-ethylmorpholine, aliphatic amines and many zwitterionic acids. For example, at concentrations of 0.05M, BES, Bicine, Cholamine, HEPES, MES, PIPES, TES and Tricine have negligible absorption above 240 nm and can be used successfully at 230 nm. ACES absorbs significantly at 230 nm and ADA is limited to above about 260 nm.

Replacement of the barbitone component of Britton and Robinson's (1931) universal buffer mixture with Tris gives buffers suitable for pH control in ultraviolet spectroscopy down to 230 nm (Davies, 1959). Details are summarized in Table 3.9. Dimethylglutarate buffers (pH 3—7) have been proposed for spectrophotometric enzyme studies (Stafford *et al.*, 1955). Succinate buffers (below pH 6.1) and glycylglycine and glycine buffers (above pH 7.3) were used in the enzymic determination of purine nucleotides around 250—290 nm (Kalckar, 1947).

Buffers for spectroscopic work should not be stored in plastic containers because they may extract plasticizers which have an appreciable ultraviolet absorption.

3.5 Buffers for special applications
3.5.1 Volatile buffers.

The pH range 2—12 can be spanned by buffers using volatile components as listed in Table 3.10. The ease with which they can be removed by simple evaporation makes them useful in high voltage electrophoresis and preparative ion-exchange chromatography. Thus, in the ion-exchange separation of amino acids or peptides, fractions can be evaporated to dryness or lyophilized for use in metabolic studies without the retention of contaminating salts. Except for the buffers containing pyridine, methylpyridine or 2,4,5-trimethylpyridine, the solutions are suitable for ultraviolet spectroscopy. Dilute solutions (less than 0.1M) of triethylamine do not cause interference in the ninhydrin reaction. Ethylenediamine/acetate buffers maintain constant pH during evaporation because ethylenediamine and acetic acid have the same boiling points. Trimethylamine is less convenient to handle and more difficult to purify than is triethylamine.

3.5.2 Buffers for electrophoresis.

Table 3.11 gives a set of constant ionic strength buffers for use in electrophoresis at 2°C. The choice of suitable buffers for zone electrophoresis is important, the best buffer salts being univalent. Boundary anomalies are minimized by using buffers in which the ion with the same sign as the solute ion has a low mobility. Examples include barbiturate and cacodylate anions and the Tris cation. However, barbiturate buffers absorb strongly in the ultraviolet, including the 284 nm region used in protein estimation. Some buffers suitable for paper electrophoresis are collected in Table 3.12. For other examples see Peeters (1959). References to buffers that have been used for starch gel electrophoresis of proteins include Smithies (1959, 1962) and Ashton (1957). An extensive list of working conditions and buffers for block and gel electrophoresis is given by Bloemendale (1963). Poulik (1957) used a Tris-citrate pH 8.65 buffer (0.076M in Tris, 0.005M in citric acid) in the gel and a borate buffer in the electrode vessel to separate proteins: this is an example of a discontinuous buffer system, the use of which leads to zone sharpening.

Tris-acetic acid buffers have been proposed instead of Tris-hydrochloric acid buffers for preparative starch gel electrophoresis because the anodic solution does not drop to such low pH values (Pilz and Johann, 1966). Examples for use at 20°C are: pH 7.3, four litres of 0.4M Tris plus 750 ml 2M acetic acid; pH 8.6, 4.5 litres 0.4M Tris plus 280 ml 2M acetic acid. A recent paper describes buffers for the electro-phoresis of enzymes on polyacrylamide gel (Orr *et al.*, 1972).

3.5.3 Buffers for complexometric titration.

The stabilities of EDTA-metal complexes and of metal ions with metallochrome indicators are strongly dependent on pH, so that by pH control increased accuracy and selectivity is possible in complexometric titrations. This can be achieved by suitable use of buffers but care must be taken that the buffers are not also strong metal-chelating agents. The final pH values of the solutions are seldom critical and, if necessary, high concentrations of buffers can be used. Cost rules out the use of imidazole buffers for pH 7.0 and most of the 'Good' buffers. The insolubility of many metal phosphates prevents the use of phosphate buffers. Suitable buffers are given in Table 3.13.

3.5.4 Buffers for chromatography.

Cationic buffers covering the pH range 3.5–10.5 have been studied for the chromatography of proteins on anionic exchangers because better separations are given at pH values near to the isoelectric points of the proteins. (Semenza *et al.*, 1962). The buffers comprise piperazine, tetrakis-(2-hydroxy-ethyl)ethylenediamine, tetraethylethylenediamine, dimethyl-aminoethylamine and 2,3-dihydroxypropyl-tris(hydroxy-methyl)methylamine. These were chosen as having good water solubility, no absorption at 280 nm, good stability on storage, no reaction with proteins, no adsorption, no inhibiting effects on enzymes, and reasonable costs. Non-exchangeable buffer ions should be used for protein chromatography on ion-exchangers. Thus, Tris or a tertiary or quaternary ammonium salt should be used with anion exchangers, whereas with cation exchangers buffer anions such as acetate, phosphate, barbiturate or borate should be used. Buffer tables covering the pH range 4.8 to 10.0 are given (Semenza *et al.*, 1962) and also some universal buffers using mixtures of these bases. Tris buffers have also been used (Boman and Westlund, 1956).

Lithium citrate buffers have been successfully used in automatic amino acid chromatography on resin columns. (See for example, Perry *et al.*, 1968; Peters *et al.*, 1968).

3.5.5 Buffers for polarography.

Because the polarographic half-wave potentials of certain types of organic compounds depend strongly on the ionic strength of the solution, Elving *et al.*, (1956) have devised modified McIlvaine citrate/phosphate buffers of constant ionic strength 0.5 and 1 (by addition of potassium chloride) which cover the range pH 2.2 to 8.0 at 25°C. A set of buffers for organic polarographic analysis might comprise the following: HCl/KCl for the strongly acid region, acetate buffers, ammonia buffers, and KOH/KCl for the strongly alkaline region. Where ammonia might react with the test material, borate buffers would be preferable.

3.5.6 Buffers for proton magnetic studies.

In using p.m.r. changes to determine the pK_a values of substituent groups in peptides, and in other pH-dependent measurements of chemical shifts, it is convenient to use acids, bases and buffer species that are lacking in protons. A convenient partial set and the pH (pD) ranges covered are:

DCl or CF_3COOD	pD 0–3
Na deuteroacetate, DCl	pD 3.8–5.8
KD_2PO_4	pD 6.2–8.2
Na_2CO_3, $NaDCO_3$	pD 9.3–11.3
NaOD	pD 11–14

3.5.7 Buffers for solvent extraction.

Many natural products, pharmaceuticals and synthetic organic chemicals are weak acids or bases which, in their non-ionized forms can be extracted from aqueous solutions by suitable organic solvents whereas, if the pH is such that they are ionized, their retention in the aqueous phase is favoured. By using buffers to control the pH of the solution, and hence to govern the concentration ratio of ionized and neutral forms, it is frequently possible to improve the selectivity of such separation processes. The use of buffers for this purpose was very important in the separation of individual penicillins and in preparing synthetic antimalarials.

Extraction of metal ions as complexes with organic

reagents is complicated by the frequent inclusion of masking agents such as aminopolycarboxylic, citric and tartaric acids in the aqueous phase to form stable, water-soluble, non-extractable complexes with potentially interfering cations. This is discussed elsewhere (Perrin, 1970).

In all cases, the buffer components should be readily soluble in the aqueous phase but be essentially insoluble in the organic phase. Nor should they form insoluble or complex species with the compounds that are to be extracted. This suggests that highly polar buffers such as citrate and phosphate would ordinarily be the buffers of choice. Acetate buffers are likely to be less suitable because of the possibility of extracting acetic acid into the organic phase. With amine buffers there is the possibility that both the neutral amine and its ion-pair with chloride ion will be extractable.

3.5.8 Isotonic pharmaceutical buffers.

Some discomfort and possible injury can be avoided in using pharmaceutical solutions that are intended for parenteral injection or for application to delicate membranes of the body if these solutions are routinely adjusted to approximate isotonicity with body fluid. (At 37°C the standard is an 0.90% sodium chloride solution.) The Tables 3.14 and 3.15 summarize recent studies on isotonic buffers at 25°C and 37°C. A 1.77% boric acid solution is isotonic with 0.9% NaCl and has a pH of 4.96 at 25°C and 4.85 at 37°C (Cutie and Sciarrone, 1969).

Several buffers have been studied at 25°C in isotonic saline (solutions adjusted to $I = 0.16$ by adding NaCl). In such solutions, equimolal ratios of Tris and Tris-HCl (each 0.05 m) gave a pH of 8.225, while for Bis-tris and Bis-tris-HCl (each 0.05 m) it was 6.647. Acetic acid (0.05 m) and sodium acetate (0.05 m) gave pH 4.637, and for 0.05 m KH_2 citrate it was 3.683 (Durst, 1970). All values are within ±0.005 pH units of predictions from the NaCl-free solutions, using the Debye-Hückel equation.

3.5.9 Miscellaneous.

Piperazine buffers (pH 5.5–9.8), made up in filtered sea water, are described by Smith and Smith (1949). Acetate/ citrate buffer was added to water samples during fluoride determinations by ion electrodes to maintain constant pH and constant ionic strength and to complex any Al^{3+} or Fe^{3+} (Frant and Ross, 1968).

A nitrogen-free universal buffer for use in biological and surface chemical work has been described by Teorell and Stenhagen (1938). It comprises a stock solution made up by dissolving 8.903 g (0.05 mole) of $Na_2HPO_4 \cdot 2H_2O$, 7.00 g (0.0333 mole) of citric acid monohydrate, 3.54 g (0.0507 mole) H_3BO_3 and 243 ml of 1M NaOH, diluted to one litre with CO_2-free water. To prepare a buffer solution of stated pH at 20°C, 20 ml of stock solution is taken and x ml of 0.1M HCl (as given in Table 3.16) is added, followed by dilution to 100 ml. The ionic strength varies between 0.07 and 0.10.

Similarly, 3,6-endomethylene-1,2,3,6-tetrahydrophthalic acid (EMTA) has been proposed as a buffer near pH 7.0 devoid of nitrogen, sulphur and phosphorus for use in bacteriological systems. (Mallette, 1967).

Liquid buffers proposed for use in the Edman degradation of peptides include pyridine (Edman, 1950), allyldimethylamine (Blombäck *et al.*, 1966) 3-dimethylamino-1-propyne (Braunitzer and Schrank, 1970) and *N,N,N',N'*-tetrakis(2-hydroxypropyl)ethylenediamine (Edman and Begg, 1967).

EMTA

Table 3.1 *Values of primary standards for pH measurement*

Temp °C	KH phthalate*		$Na_2B_4O_7 \cdot 10H_2O$†	
	0.05m‡ §	0.05M ‖	0.01m‡ ¶	0.01M **
0	4.003	4.011	9.464	9.454
5	3.999	4.005	9.395	
10	3.998	4.001	9.332	9.330
15	3.999	4.000	9.276	
20	4.002	4.001	9.225	9.228
25	4.008	4.005	9.180	9.185
30	4.015	4.011	9.139	9.144
35	4.024	4.020	9.102	
40	4.035	4.031	9.068	9.076
45	4.047	4.045	9.038	
50	4.060	4.061	9.011	9.020
55	4.075	4.080	8.985	
60	4.091	4.091	8.962	8.975
65	4.108	4.105	8.942	
70	4.126	4.121	8.921	
75	4.145	4.140	8.903	
80	4.164	4.161	8.885	
85	4.184	4.184	8.867	
90	4.205	4.211	8.850	
95	4.227	4.240	8.833	

*$\Delta pH_{1/2}$ = 0.052
†$\Delta pH_{1/2}$ = 0.01
‡U.S. National Bureau of Standards (Bates, 1962)
§ Molal scale. Made by dissolving 10.12 g of $KHC_8H_4O_4$ in 1 kg of distilled water.
‖British Standard (1961). (10.21 g $KHC_8H_4O_4$ in 1 litre at 25°C).
¶ Made by dissolving 3.80 g of $Na_2B_4O_7 \cdot 10H_2O$ in 1 kg of distilled water.
**Alner et al. (1967). (3.81 g of $Na_2B_4O_7 \cdot 10H_2O$ in 1 litre at 25°C).

Table 3.2 *Some alternative buffers for pH standardization**

Solution	pH at given temperature (°C)							
	0°	10°	20°	25°	30°	40°	50°	60°
0.1 m K tetroxalate	—	—	1.475	1.482	1.486	1.500	1.514	1.524
0.05 m K tetroxalate	1.631	1.638	1.647	1.650	1.651	1.654	1.666	1.670
0.01M HCl, 0.09M KCl				2.07†		2.08(38°)†		
0.034 m KH tartrate (sat. at 25°C)				3.559 3.557‡	3.554	3.550 3.548(38°)‡	3.553 3.549‡	3.564
0.05 m KH₂ citrate	3.863§	3.820§	3.788§	3.776§‖	3.776§	3.753§	3.749§	
0.1M acetic acid, 0.1M Na acetate	4.667	4.655 4.65(12°)¶	4.645	4.646 4.64†	4.648	4.659 4.65(38°)†	4.673	4.695
0.01M acetic acid, 0.01M Na acetate	4.732	4.770 4.71(12°)¶	4.713	4.716¶ 4.70¶	4.720	4.734 4.72(38°)¶	4.752	4.799
0.025 m KH₂PO₄, 0.025 m Na₂ succinate				6.109**††				
0.025 m KH₂PO₄, 0.025 m Na₂HPO₄	6.936	6.916 6.940(15°)‡	6.874	6.858 6.865‡	6.857	6.832 6.840(38°)‡	6.825 6.833‡	6.827
0.05 m sodium borate	9.479	9.347	9.236	9.183	9.142	9.056	8.994	8.943
0.025 m NaHCO₃, 0.025 m Na₂CO₃	10.317§	10.179§	10.062§	10.012§‡‡	9.966§	9.889§	9.828§	
0.0211 m Ca(OH)₂, (sat. at 20°)	13.364	12.964	12.604	12.435	12.273	11.966	11.687	11.435

*Alner et al. (1967)
† Hitchcock & Taylor (1938)
‡ MacInnes et al. (1938)
§ Staples & Bates (1969)
‖ $\Delta pH_{\frac{1}{2}} = 0.024$
¶ Bates (1962)
** $\Delta pH_{\frac{1}{2}} = 0.071$
†† Paabo et al. (1963)
‡‡ $\Delta pH_{\frac{1}{2}} = 0.079$

Table 3.2 *(Cont.)*

0.1 m K tetroxalate	$= 0.09841M$ at $25°C$
	$= 25.0148$ g $KH_3(C_2O_4)_2 \cdot 2H_2O/l$
0.05 m K tetroxalate	$= 0.04952M$ at $25°C$
	$= 12.5875$ g $KH_3(C_2O_4)_2 \cdot 2H_2O/l$
0.05 m KH_2 citrate	$= 0.04958M$ at $25°C$
	$= 11.41$ g $KH_2C_6H_5O_7/l$
0.025 m KH_2PO_4	$= 0.02498M$ at $25°C$
	$= 3.3995$ g KH_2PO_4/l
0.025 m Na_2HPO_4	$= 0.02498M$ at $25°C$
	$= 3.5463$ g Na_2HPO_4/l
0.05 m sodium borate	$= 0.0491M$ at $25°C$
	$= 18.7282$ g $Na_2B_4O_7 \cdot 10H_2O/l$
0.025 m $NaHCO_3$	$= 0.02492M$ at $25°C$
	$= 2.092$ g $NaHCO_3/l$
0.025 m Na_2CO_3	$= 0.02492M$ at $25°C$
	$= 2.640$ g Na_2CO_3/l

Table 3.3 *NBS phosphate buffers as pH standards**

Temp. ($°C$)	Soln. I†	Soln. II‡	Temp. ($°C$)	Soln. I†	Soln. II‡
0	6.984	7.534	50	6.833	7.367
5	6.951	7.500	55	6.834	
10	6.923	7.472	60	6.836	
15	6.900	7.448	65	6.840	
20	6.881	7.429	70	6.845	
25	6.865 §	7.413 ‖	75	6.852	
30	6.853	7.400	80	6.859	
35	6.844	7.389	85	6.868	
38	6.840	7.384	90	6.877	
40	6.838	7.380	95	6.886	
45	6.834	7.373			

*Bates (1962)
†0.025M KH_2PO_4, (3.388 g)
and 0.025M Na_2HPO_4 (3.533 g) in 1 l of solution at $25°C$)
‡0.00869M KH_2PO_4, (1.179 g),
and 0.0304M Na_2HPO_4, (4.302 g), in 1 l of solution at $25°C$).
§ $\Delta pH_{1/2} = 0.080$
‖ $\Delta pH_{1/2} = 0.07$

Table 3.4 *Bicarbonate/CO_2 buffers* at $37°C$*

(For 5% CO_2 in gas phase, and an atmospheric pressure of 725–760 mm Hg)

pH	Conc. NaHCO₃ (mM)	pH	Conc. NaHCO₃ (mM)
6.0	0.586	7.2	9.29
6.2	0.929	7.4	14.7
6.4	1.47	7.6	23.3
6.6	2.33	7.8	37.0
6.8	3.70	8.0	58.6
7.0	5.86		

*Dawson *et al.*, (1969)

Table 3.5 pK_a' *Data for some zwitterionic buffers*

Name *	pK_a' at $20°C$ (and 0.1M)	Δp$K_a'/°C$	Sat. soln. ($0°C$) (M)
MES	6.15†	−0.011	0.65
ADA	6.62†	−0.011	−
PIPES	6.82†	−0.009	−
ACES	6.88†	−0.020	0.22
BES	7.17†	−0.016	3.2
MOPS	7.20	−	−
TES	7.50†	−0.020	2.6
HEPES	7.55†	−0.014	2.25
EPPS (HEPPS)	8.00	−	−
Tricine	8.15†	−0.021	0.8
Bicine	8.35†	−0.018	1.1
TAPS	8.4	−	−

*For full names, see Appendix III
†Good *et al.* (1966)

Table 3.6 *Buffer compositions of 0.05M HEPES solutions at 20°C and I = 0.1.*

(By titration of 11.92 g HEPES with 1M NaOH, and 1M NaCl and dilution to 1 litre)

pH	ml NaOH	ml NaCl	pH	ml NaOH	ml NaCl
6.6	5.0	95.0	7.6	26.4	73.6
6.7	6.2	93.8	7.7	29.3	70.7
6.8	7.5	92.5	7.8	32.0	68.0
6.9	9.1	90.9	7.9	34.6	65.4
7.0	11.0	89.0	8.0	36.9	63.1
7.1	13.1	86.9	8.1	39.0	61.0
7.2	15.4	84.6	8.2	40.9	59.1
7.3	18.0	82.0	8.3	42.5	57.5
7.4	20.7	79.3	8.4	43.8	56.2
7.5	23.6	76.4	8.5	45.0	55.0

Table 3.7 *For Table 3.6 to apply to other HEPES-type buffer substances make the following pH corrections:*

For		pH units	For		pH units
ACES	subtract	0.67	MOPS	subtract	0.35
ADA	subtract	0.93	TAPS	add	0.85
BES	subtract	0.38	TES	subtract	0.05
Bicine	add	0.80	Tricine	add	0.60
MES	subtract	1.40			

Table 3.8 Buffers for spectrophotometry of aqueous solutions (20°C).*

(I = 0.01, final volume 100 ml)

Chloroacetic Acid			Formic Acid			Acetic Acid		
pH	0.1M Acid (ml)	0.1M KOH (ml)	pH	0.1M Acid (ml)	0.1M KOH (ml)	pH	0.1M Acid (ml)	0.1M KOH (ml)
2.20	50.99	2.76	3.20	41.34	9.30	4.10	51.26	9.91
2.30	42.18	4.31	3.30	34.99	9.45	4.20	43.03	9.93
2.40	35.38	5.51	3.40	30.28	9.55	4.30	36.17	9.94
2.50	29.98	6.46	3.50	25.88	9.65	4.40	30.87	9.95
2.60	25.82	7.19	3.60	22.62	9.72	4.50	26.67	9.96
2.70	22.46	7.79	3.70	20.21	9.78	4.60	23.14	9.97
2.80	19.94	8.23	3.80	18.05	9.82	4.70	20.58	9.98
2.90	17.85	8.60	3.90	16.36	9.86	4.80	18.38	9.98
3.00	16.21	8.89	4.00	15.03	9.89	4.90	16.69	9.99
3.10	14.95	9.12	4.10	14.00	9.91	5.00	15.34	9.99
3.20	13.89	9.30	4.20	13.21	9.93	5.10	14.28	9.99
3.30	13.11	9.45	4.30	12.54	9.94	5.20	13.42	9.99
3.40	12.46	9.55	4.40	12.05	9.95	5.30	12.78	9.99

Phosphoric Acid

pH	0.02M KH$_2$PO$_4$ (ml)	0.01M Na$_2$HPO$_4$ (ml)
6.40	30.21	13.19
6.50	27.45	15.05
6.60	24.60	16.94
6.70	21.66	18.90
6.80	18.97	20.68
6.90	16.40	22.41
7.00	13.96	24.04
7.10	11.72	25.52
7.20	9.73	26.85
7.30	7.99	28.00
7.40	6.54	28.97
7.50	5.32	29.78
7.60	4.27	30.49
7.70	3.41	31.05

Succinic Acid

pH	0.1M acid (ml)	0.05M KOH (ml)	pH	0.02M Acid (ml)	0.05M KOH (ml)
3.60	45.95	19.21	4.80	46.04	17.57
3.70	36.93	19.27	4.90	41.62	17.15
3.80	31.84	19.27	5.00	38.12	16.80
3.90	27.12	19.24	5.10	35.01	16.42
4.00	23.29	19.21	5.20	32.15	16.02
4.10	19.72	19.11	5.30	29.88	15.68
4.20	17.50	19.00	5.40	27.70	15.34
4.30	15.37	18.84	5.50	25.89	15.05
4.40	13.73	18.67	5.60	24.33	14.78
4.50	12.24	18.42	5.70	22.94	14.53
4.60	11.06	18.15	5.80	21.95	14.35
4.70	10.12	17.90	5.90	20.82	14.13
			6.00	20.07	14.00
			6.10	19.41	13.88
			6.20	18.87	13.77

Table 3.8 *(Cont.) Buffers for spectrophotometry of aqueous solutions* (20°C)* (I = 0.01, final volume 100 ml)

Boric Acid

pH	0.025M Na$_2$B$_4$O$_7$ (ml)	0.1M H$_3$BO$_3$ (ml)
8.50	20.00	38.62
8.60	20.00	29.89
8.70	20.00	22.97
8.80	20.00	16.62
8.90	20.00	11.21
9.00	20.00	6.89
9.10	20.00	3.07
9.20	20.00	0.60

pH	0.025M Na$_2$B$_4$O$_7$ (ml)	0.01M KOH (ml)
9.30	18.30	8.56
9.40	16.47	17.62
9.50	15.01	24.97
9.60	13.74	31.25
9.70	12.94	35.29

Carbonic Acid

pH	0.02M NaHCO$_3$ (ml)	0.01M Na$_2$CO$_3$ (ml)
9.60	29.64	13.64
9.70	26.26	15.92
9.80	23.05	18.08
9.90	20.32	19.80
10.00	17.41	21.89
10.10	14.47	23.92
10.20	11.76	25.79
10.30	9.32	27.51
10.40	7.35	28.95
10.50	5.32	30.43
10.60	3.61	31.77

Trishydroxymethyl-aminomethane

pH	0.1M Tris (ml)	0.1M HCl (ml)
7.70	12.97	10.00
7.80	13.68	10.00
7.90	14.67	10.00
8.00	15.84	10.00
8.10	17.28	10.00
8.20	19.14	10.00
8.30	21.52	10.00
8.40	24.44	10.00
8.50	28.18	10.00
8.60	32.90	10.00
8.70	38.87	10.00
8.80	46.41	10.00
8.90	55.22	10.00

Citric Acid			Butylamine					
pH	0.002M Acid (ml)	0.01M Na$_3$ Citrate (ml)	pH	0.2M Free Base (ml)	0.1M HCl (ml)	pH	0.02M Free Base (ml)	0.1M HCl (ml)
5.80	33.12	19.78	10.30	6.45	9.85	11.50	28.02	7.57
5.90	28.88	19.41	10.40	6.83	9.81	11.60	33.97	6.94
6.00	25.10	19.07	10.50	7.30	9.76			
6.10	21.45	18.75	10.60	7.90	9.69			
6.20	18.26	18.46	10.70	8.65	9.61			
6.30	15.27	18.17	10.80	9.59	9.51			
6.40	12.73	17.92	10.90	10.78	9.39			
6.50	10.47	17.70	11.00	12.28	9.23			
6.60	8.58	17.51	11.10	14.16	9.03			
6.70	6.99	17.37	11.20	16.53	8.78			
			11.30	19.52	8.46			
			11.40	23.28	8.07			

*Perrin (1963).

Table 3.9 *Universal buffer (pH 2–12) for u.v. spectrophotometry**†

(50 ml of a solution 0.1M in citric acid, (21.01 g/l); 0.1M in KH_2PO_4, (13.61 g/l); 0.1M in sodium tetraborate, (19.07 g/l); 0.1M in Tris, (12.11 g/l); 0.1M in KCl, (7.46 g/l), to which is added x ml 0.4M HCl or 0.4M NaOH, followed by dilution to 200 ml)

pH (25°C)	pH (37°C)	ml 0.4M HCl	pH (25°C)	pH (37°C)	ml 0.4M HCl
2.00	2.02	34.8	3.60	3.58	14.0
2.20	2.21	30.4	3.80	3.78	12.0
2.40	2.40	26.6	4.00	3.98	10.0
2.60	2.60	23.8	4.20	4.18	7.8
2.80	2.79	21.6	4.40	4.38	5.6
3.00	2.98	19.6	4.60	4.58	3.6
3.20	3.18	17.6	4.80	4.78	1.6
3.40	3.38	15.8			

pH (25°C)	pH (37°C)	ml 0.4M NaOH	pH (25°C)	pH (37°C)	ml 0.4M NaOH
5.00	4.99	0.4	8.60	8.47	39.8
5.20	5.19	2.8	8.80	8.66	43.4
5.40	5.39	5.0	9.00	8.83	46.2
5.60	5.58	7.4	9.20	9.03	49.0
5.80	5.78	9.4	9.40	9.22	52.0
6.00	5.98	11.4	9.60	9.42	54.6
6.20	6.18	13.4	9.80	9.61	56.8
6.40	6.37	15.6	10.00	9.81	59.0
6.60	6.56	17.8	10.20	10.00	60.4
6.80	6.76	20.2	10.40	10.20	61.6
7.00	6.96	22.4	10.60	10.38	62.8
7.20	7.15	24.0	10.80	10.58	64.0
7.40	7.34	26.6	11.00	10.77	65.6
7.60	7.53	28.6	11.20	10.97	67.0
7.80	7.72	30.8	11.40	11.17	68.8
8.00	7.91	33.2	11.60	11.36	71.0
8.20	8.10	35.6	11.80	11.56	73.8
8.40	8.29	37.6	12.00	11.76	77.2

*Davies (1959)

†I varies from 0.09 (at pH 3) to 0.33 (at pH 12).

Table 3.10 *Types of systems for use as volatile buffers*

System	pH Range
87 ml Glacial acetic acid + 25 ml 88% HCOOH in 1 l	1.9
25 ml 88% HCOOH in 1 l	2.1
Pyridine-formic acid	2.3−3.5
Trimethylamine-formic acid	3.0−5.0
Triethylamine-formic (or acetic) acid	3−6
5 ml Pyridine + 100 ml glacial acetic acid in 1 l	3.1
5 ml Pyridine + 50 ml glacial acetic acid in 1 l	3.5
Trimethylamine-acetic acid	4.0−6.0
25 ml Pyridine + 25 ml glacial acetic acid in 1 l	4.7
Collidine-acetic acid	5.5−7.0
100 ml Pyridine + 4 ml glacial acetic acid in 1 l	6.5
Triethanolamine-HCl	6.8−8.8
Ammonia-formic (or acetic) acid	7.0−10.0
Trimethylamine-CO_2	7−12
Triethylamine-CO_2	7−12
0.3M NH_4HCO_3	7.9
Ammonium carbonate-ammonia	8.0−9.5
Ethanolamine-HCl	8.5−10.5
20 g $(NH_4)_2CO_3$ in 1 l	8.9

Table 3.11 *Constant ionic strength buffers for electrophoresis (2°C)**

(Final volume 1 l For I = 0.1, add 16 ml 5M NaCl; for I = 0.2, add 36 ml 5M NaCl.)

pH	Glycine, NaCl (each 1M) ml	HCl 1M ml	NaOAc† 1M ml	HOAc 1.75M ml	Na$_2$HPO$_4$ 0.5M ml	NaH$_2$PO$_4$ 2M ml	Na Barb‡ 0.5M ml	NaOH 1M ml
2.0	5.3	14.7						
2.5	11.4	8.6						
3.0	15.8	4.2						
3.5	18.3	1.7						
4.0			20.0	33.7				
4.5			20.0	11.5				
5.0			20.0	3.7				
5.5			20.0	1.2				
6.0					4.6	6.6		
6.5					8.4	3.7		
7.0					11.35	1.6		
7.5					12.15	0.5		
8.0		10.4					40.0	
8.5		5.3					40.0	
9.0		2.0					40.0	
9.5	17.3							2.7
10.0	14.4							5.6
10.5	11.6							8.4
11.0	9.8							10.2
11.5	8.8							11.2
12.0	7.6							12.4

*Miller & Golder (1950)   †OAc = acetate   ‡Barb = diethylbarbiturate

Table 3.12 *Buffers for paper electrophoresis of proteins*

pH	Ionic strength	Composition (per litre)
4.4	0.2	9.44 g Na_2HPO_4 + 10.3 g citric acid
4.5	0.1	3.5 g NaCl + 3.28 g NaOAc, adjusted to pH 4.5 with 1M HCl
5.9*		5.10 g KH phthalate + 0.86 g NaOH
6.5	0.1	3.11 g KH_2PO_4 + 1.49 g Na_2HPO_4
7.8	0.12	0.294 g $NaH_2PO_4 \cdot 2H_2O$ + 3.25 g Na_2HPO_4
8.6†	0.1	3.68 g diethylbarbituric acid + 20.6 g sodium diethylbarbiturate
8.6		9.78 g sodium diethylbarbiturate + 6.47 g NaOAc + 60 ml 0.1M HCl
8.6		8.8 g sodium borate + 4.65 g H_3BO_3
8.6		8.76 g sodium diethylbarbiturate + 1.38 g diethylbarbituric acid + 0.384 g calcium lactate
8.9‡		60.5 g Tris, 6.0 g EDTA + 4.6 g H_3BO_3
9.0		7.63 g sodium borate + 0.62 g H_3BO_3
10.0§		7.44 g H_3BO_3 + 100 ml 1M NaOH

*amino acids
†pH 8.6 buffers are for serum proteins
‡for serum proteins (Aronsson and Grönwall, 1957)
§ carbohydrates

Table 3.13 *Buffers for complexometric titrations*

pH	
2–4	Glycine, *p*-chloroaniline, monochloracetic acid
3.7	Chloracetic acid – sodium acetate (equimolar mixture)
4–6.5	Acetic acid (e.g., 500 ml 0.2M sodium acetate + 500 ml 0.2M acetic acid gives pH 4.6)
5	Acetic acid-pyridine (equimolar mixture)
5–6	Hexamethylenetetramine (urotropine) (e.g., 400 g hexamethylenetetramine, 100 ml conc. HCl and 1 l water gives pH 5.4)
6.5–8	Triethanolamine* or possibly maleate* (pH 6–7)
8–11	Ammonia† (e.g., 70 g NH_4Cl + 570 ml ammonia (S.G. 0.90) diluted to 1 l gives pH ~ 10)
12.5	Diethylamine‡ (82 g diethylamine hydrochloride in 800 ml of distilled water, add 80 ml 50% (w/v) NaOH and dilute to 1 l)

*Keep concentrations low, because this is a strong complexing agent.
†Glycine unsuitable in alkaline solution because of its complexing ability at high pH.
‡Store in a stoppered polythene bottle.

Table 3.14 *pH Values of isotonic Palitzsch* buffers†*

(x ml 0.05M sodium tetraborate ($Na_2B_4O_7 \cdot 10H_2O$, 19.108 g l^{-1}), $100 - x$ ml of 0.2M boric acid (12.404 g l^{-1}), the final mixture also containing NaCl to adjust the tonicity to 0.90 ± 0.01 at $37°C$)

x	mg NaCl 100 ml^{-1}	pH, $25°C$	pH, $37°C$
3	270	6.82	6.80
6	270	7.16	7.13
10	270	7.44	7.41
15	260	7.67	7.64
20	260	7.85	7.81
25	250	7.98	7.95
30	240	8.12	8.11
35	230	8.23	8.20
45	210	8.43	8.39
55	190	8.60	8.54
60	180	8.67	8.62
70	140	8.81	8.76
80	110	8.95	8.88
90	70	9.08	8.99

*Palitzsch (1915)
†Cutie & Sciarrone (1969)

Table 3.15 *pH Values of isotonic Sörensen* buffers†*

(x ml Na_2HPO_4 (9.470 g l^{-1}), $100 - x$ ml $NaH_2PO_4 \cdot H_2O$ (9.208 g l^{-1})‡, the final mixture also containing NaCl to adjust the tonicity to 0.92 at $37°C$)

x	mg NaCl 100 ml^{-1}	pH, $25°C$	pH, $37°C$
10	520	5.76	5.74
20	510	6.12	6.10
30	500	6.35	6.33
40	490	6.54	6.53
50	480	6.72	6.71
60	460	6.89	6.88
70	450	7.09	7.08
80	440	7.33	7.32
90	430	7.70	7.69
95	420	8.07	8.05

*Sörensen (1909)
†Cutie & Sciarrone (1969)
‡or 10.41 g l^{-1} $NaH_2PO_4 \cdot 2H_2O$

Table 3.16 Volume of 0.1M HCl to be included in Teorell and Stenhagen's buffer mixture* to give required pH at 20°C.

pH	0.00 ml	0.10 ml	0.20 ml	0.30 ml	0.40 ml	0.50 ml	0.60 ml	0.70 ml	0.80 ml	0.90 ml
2		72.10	69.25	66.87	64.90	63.25	61.77	60.48	59.29	58.29
3	57.49	56.76	56.05	55.42	54.83	54.28	53.72	53.17	52.61	52.07
4	51.52	51.00	50.46	49.92	49.40	48.88	48.35	47.81	47.28	46.72
5	46.18	45.64	45.10	44.54	43.99	43.40	42.77	42.15	41.51	40.89
6	40.28	39.66	39.02	38.31	37.54	36.73	36.02	35.36	34.72	34.13
7	33.51	32.97	32.46	31.90	31.36	30.82	30.33	29.88	29.45	29.06
8	28.70	28.44	28.20	27.91	27.56	27.20	26.83	26.34	25.77	25.12
9	24.48	23.82	23.21	22.60	21.95	21.32	20.71	20.13	19.60	19.10
10	18.65	18.24	17.84	17.51	17.20	16.92	16.68	16.35	15.98	15.56
11	15.09	14.59	13.92	13.08	12.09	10.75				

*Teorell & Stenhagen (1938); see page 38

Chapter Four

Practical Limitations in the use of Buffers

4.1 Chemical problems

The suitability of a buffer system for any particular application depends on many factors, the first of which is the pK_a value of the buffer acid or base. For a buffer to be effective, its pH must be within the range $pK_a \pm 1$, or preferably, within $pK_a \pm 0.5$. The former scarcity of buffer substances with a pH range 6 to 8 frequently led to the use of buffers such as phosphate or Tris in pH regions where they had little buffer capacity.

Not only must the buffer species be appreciably soluble in water, but it is also important that they do not react with ions or molecules present in the solution. For example, phosphate and pyrophosphate buffers are unsuitable if a solution contains calcium or certain other di- and trivalent cations which form insoluble phosphates, or if reaction progress is to be followed by the change in phosphate content. Similarly, carbonate buffers can precipitate calcium ion.

Borate buffers should not be used in the presence of polyols, including carbohydrates and their derivatives, with which they may form chelate compounds: they react in this way with many respiratory intermediates. Use of borate buffers in gel electrophoresis of proteins can result in spreading of the zones if fructose, ribose, sorbitol, catechol or other appropriate polyols are present to form borate complexes (Lerch and Stegemann, 1969). The increased solubility of adrenalin in borate buffers is due to complex formation as is also the improved separation of sugar phosphates chromatographically in the presence of borate buffers. Likewise, complexation of carbohydrates in germanate or borate buffers is the basis of a suggested method for their paper electrophoretic separation (Lindberg and Swan, 1960). With 0.1M borate concentrations, complex formation by polyols can lower the pH by up to three or four pH units. Borate buffers also have a high bacteriocidal effect.

A difficulty in using bicarbonate/CO_2 buffers is that pH constancy requires the use of closed systems equilibrated with a controlled level of CO_2 in the gas phase. Otherwise, loss of CO_2 from the solution can lead to a progressive and undesirable rise in pH.

(Buffers with pH ranges similar to phosphate and bicarbonate ions, that might be suitable as alternatives, include PIPES, ACES, Cholamine, BES, MOPS, TES, and HEPES. Orthophosphite ($pK_2' = 6.5$) has also been proposed as a replacement (pH range 5.5 to 7.5) for phosphate buffers in biological studies (Robertson and Boyer, 1956), because of non-interference in enzymic reactions of phosphate, relative non-toxicity to yeast and animal cells, and improved solubility of calcium and magnesium salts. Phosphite solutions are not readily oxidized by oxygen. Borate might be replaced by 2-amino-2-methyl (or ethyl)-1,3-propanediol, diethanolamine, ethanolamine,4-aminopyridine or such zwitterionic buffers as 2-amino-ethylsulphonic acid, serine or CHES.)

On prolonged storage, borax is likely to lose some of its water of crystallization unless the container is tightly stoppered. (Alternatively, the correct degree of hydration of borax can be maintained as described in Chapter 8). This is not serious when borax is used, on its own, as a buffer standard but it can lead to significant errors in buffers when its strength as an acid or a base is important. The buffers in Tables 10.37a and 10.41 are examples and their pH values should always be checked by pH meter. This is also a disadvantage of Kolthoff's (1925, 1932) borax-phosphate and borax-succinate buffers which could be overcome by reformulating in terms of boric acid, sodium succinate and phosphate.

The storage of alkalis and alkaline buffer solutions presents problems because of the avidity with which CO_2 is absorbed and because of the slow attack by the reagents on the bottles in which they are stored. With phosphate solutions this can lead to the deposition of calcium phosphate.

Where the buffer has metal-complexing ability, problems may arise, particularly if heavy metal ions are present. Competition between metal ions and protons for attachment to buffer species can lead to a lowering of pH. Thus, the pH of phosphate and citrate buffers is markedly reduced by the addition of calcium ions (Davies and Hoyle, 1953, 1955). This effect is also to be expected with heavy metal ions and amines or amino acids. In addition, copper ions can displace protons from peptide groups.

A practical limitation of a different kind arises in measurements when the pH is less than 3 or more than 11. The liquid junction potential of the common pH cell shows a great variability under these conditions, preventing, for example, the use of potassium tetroxalate (pH = 1.675 at 20°C) as a primary pH standard.

Other factors which need to be allowed for are the change in pH with dilution of a buffer and, more importantly, the effect of temperature on the pK_a of a buffer. Tris has a pK_a of 8.85 at 0°C, 8.06 at 25°C and 7.72 at 37°C, so that the pH of a Tris buffer can fall by more than 1 pH unit in warming from 0°C to 37°C. Comparable effects are found for other cationic buffers such as the aliphatic amines.

The cost of buffer substances may be an important consideration, especially where large quantities of material are involved. Another limitation may be the difficulty in obtaining the buffer substance in an adequate degree of purity: this is a problem with triethanolamine.

Mould growth may occur in buffers in the pH range 3—11 if organic acids or bases are present. Potassium hydrogen tartrate is particularly susceptible. When mould growth is observed the buffer should be discarded. Alternatively, the useful lives of solutions can be prolonged by the initial addition of a small amount of a suitable preservative, such as a few crystals of thymol for potassium hydrogen tartrate or phthalate solutions. (A saturated solution of thymol at room temperature contains about 1 g l^{-1}). Diethyl pyrocarbonate (0.1 ml in 1 ml of ethanol, added to 1 litre of buffer) has been claimed to be better than pentachlorophenol or octanoic acid as a mould growth inhibitor in citric acid buffers for automatic amino acid analysis (Maravalhas, 1969). An alternative means of reducing microbial contamination of such buffers might be to pass the buffer solution through an appropriate filter column. Some commercially available solutions of Sörensen's and Clark and Lubs' buffers for the range pH 4 to 9 contain 20 ppm of $K_2 HgI_4$ as a preservative.

The apparent failure of moulds to metabolize α-hydroxyisobutyric acid has led to the suggestion that this acid might be useful for buffers (pH 2.3–5.0); a table is available (Tobie and Ayres, 1945).

4.2 Biological effects

Buffer species can exert effects on biological systems in three main ways. They may specifically stimulate or depress enzyme activity. They may interfere or react with substrates, inhibitors or cofactors. Non-specifically they may exert effects because of their ionic strength. In general, buffer concentrations should be kept as low as possible, having regard for the need to maintain pH constancy, and the medium should be adjusted by adding appropriate inorganic and organic ions to simulate physiological conditions. Metal-ion complexation by buffers can also be important.

Some common examples of enzymes inhibited by phosphate ions include carboxypeptidase, fumarase, urease, phosphoglucomutase, carboxylase, arylsulphatase and muscle deaminase (for the deamination of adenylic acid). Frequently this inhibition is due to competition of the phosphate with substrates containing phosphate groups or to complex formation with a metal ion essential for the enzyme activity.

Barbiturate buffers can uncouple oxidative phosphorylation and are also limited by the low solubility of diethylbarbituric acid. Glycine can serve as a substrate in a number of fermentation processes. The presence of nitrogen in barbiturate, glycine and zwitterionic buffers complicates the study of proteins and nitrogenous substances.

The sensitivity of biological systems to the buffers used in their investigation has frequently been shown, for example by Stinson and Spencer (1968) who evaluated the effects of five buffers on respiratory parameters of isolated mitochondria. In this case TES was superior in most respects whereas Tris behaved poorly. The use of phenolic buffers may not be possible in physiological media because of the antiseptic properties of the phenols. Imidazole is too reactive and unstable to be satisfactory as a biological buffer. MOPS and MES undergo some decomposition when they are autoclaved in the presence of glucose. In the Hill reaction Bicine is unsuitable because it is slowly oxidized by ferricyanide. Bicine and Tricine are rapidly photooxidized by flavin mononucleotides or flavoproteins in visible light, leading to reduction or to a rapid oxygen uptake: TES and MES react somewhat more slowly (Yamazaki and Tolbert, 1970). Bicine is unsuitable in glucose determinations by the Hagedorn-Jensen method because it is oxidized by heating with alkaline potassium ferricyanide. EPPS (HEPPS) and HEPES interfere in the Folin protein assay (but not in the biuret test). Buffers such as HEPES, EPPS and Bicine give false positive colours in the Lowry method of protein determination: others interfere to a much smaller extent but some buffers decrease the extent of colour formation. However, MES and TES appear to be satisfactory (Gregory and Sajdera, 1970).

The possibility always exists that any given buffer may exert undesirable effects in the biochemical or biological system being studied. To minimize the danger of selecting an unsuitable buffer it is desirable, in the early stages of an investigation, to carry out replicate studies using structurally different types of buffers.

Although Tris has been a major biochemical buffer for many years, partly because it is relatively inexpensive and readily available in a highly purified form, it has disadvantages. These include its reactivity as a primary amine and its appreciable solubility in organic solvents which leads to its accumulation in the biological phases of reaction systems. Thus, Tris buffer displaces the electron transport-, and phosphorylation-, pH rate profiles for chloroplasts by almost a pH unit when compared with a number of other buffers. It also inhibits isocitrate dehydrogenase of pea mitochondria whereas HEPES does not.

Metal-binding by buffer species may be desirable or not, depending on whether the reaction being studied is metal-ion mediated. Inactivation of enzymes by heavy metal ions such as $Ag(I)$, $Cu(II)$, $Hg(II)$ and $Pb(II)$ can often be avoided by adding suitable chelating agents such as EDTA. In many cases it might be sufficient to use glycine, or some other suitable material, both as buffer and chelating agent. The strong copper-binding ability of ADA, Bicine and Tricine might thus be advantageous. Conversely, MES and HEPES are sufficiently weak chelating agents that they are suitable as buffers for studying copper(II)-binding by amino acids and peptides. Stability constants of calcium and magnesium with ADA are too large for ADA to be satisfactory in systems such as ATPase, alkaline phosphatase or phospholipase that are activated by these metal ions.

4.3 Influence on chemical reactions

Rates of chemical processes may be affected by general acid-base catalysis. For example, hydrolysis of aliphatic esters of halogen-containing carboxylic acids is catalysed by buffer anions, the rate increasing linearly with the buffer concentration. Usually, however, catalysis by buffer species is less important than catalysis by hydrogen or hydroxide ions. Amines can also catalyse ester hydrolysis: at high pH values Tris reacts with *p*-nitrophenylacetate to form an *O*-acetyl Tris derivative.

The decarboxylation of ascorbic acid is catalysed by

phosphate, oxalate and boric acid. Buffers catalyse the mutarotation of α- and β-glucose, and the epimerization of tetracycline is catalysed by phosphate buffer. Borate buffers inhibit the hydrolysis of riboflavine but this is due to borate complex formation with the ribosyl group. Direct reaction between buffer and substances present can also occur, as when Tris reacts with glyoxal to form a *N*-hemiacetal.

These examples are by no means exhaustive. For further discussion see Gensch (1967).

Chapter Five

New pH-Buffer Tables and Systems

Although available buffer tables are extensive, they may fail to meet particular requirements, such as a specified ionic strength or a nominated buffer species. Where pK_a values of the buffer substances are known the compositions of the solutions can be calculated as indicated below. Considerations governing the design of new types of buffer systems are also discussed.

5.1 On calculating buffer composition tables
5.1.1 Buffers of constant ionic strength. No added electrolyte.

For a buffer system

$$HA^{(n-1)-} \rightleftharpoons H^+ + A^{n-}$$

in which the counter ions (sodium, potassium, etc.) are monovalent, the ionic strength is

$$I = \tfrac{1}{2}n(n-1)[HA^{(n-1)-}] + \tfrac{1}{2}n(n+1)[A^{n-}] \qquad (5.1)$$

if $[H^+]$ is small. For $NaHCO_3/Na_2CO_3$, $n = 2$, so that $I = [NaHCO_3] + 3[Na_2CO_3]$. In the system CH_3COOH/CH_3COONa, $n = 1$ and $I = [CH_3COONa]$. Thus for a buffer comprising a weak monobasic acid and its alkali metal salt the ionic strength is equal to the concentration of this salt. Similarly for a weak base (such as ammonia or Tris) and its salt with an acid such as HCl, HNO_3 or $HClO_4$ the ionic strength is again equal to the concentration of the salt.

Having calculated the ionic strength it becomes possible to convert a thermodynamic pK_a given in Appendix III to a practical pK_a' for the given ionic strength. This is done by using a convenient form of the Debye-Hückel equation, such as that due to Davies (1938)

$$pK_a' = K_a - (2n-1)\{0.5I^{1/2}/(1 + I^{1/2}) - 0.1I\} \qquad (5.2)$$

For the loss of a proton from BH^{n+} the corresponding relation is

$$pK_a' = pK_a + (2n - 1)\{0.5I^{\frac{1}{2}}/(1 + I^{\frac{1}{2}}) - 0.1I\} \qquad (5.3)$$

Values of the term within braces can be read directly from Table 2.3.

The concentration ratio for the buffer species can now be obtained by using the rearranged forms of the Henderson-Hasselbalch equation,

$$\text{antilog}(pH - pK_a') = [A^{n-}]/[HA^{(n-1)-}] = x \qquad (5.4)$$

$$\text{antilog}(pH - pK_a') = [B^{(n-1)+}]/[BH^{n+}] = x \qquad (5.5)$$

Substitution into expressions for the ionic strength then gives

for $\quad HA^-/A^{2-}$, $\quad [HA^-] = I/(1 + 3x)$, $[A^{2-}] = xI/(1 + 3x) \qquad (5.6)$

for $\quad HA/A^-$, $\qquad\quad [HA] = I/x$, $[A^-] = I \qquad (5.7)$

for $\quad BH^+/B$, $\qquad\quad [BH^+] = I$, $[B] = xI \qquad (5.8)$

These expressions enable buffer compositions at given ionic strengths to be calculated as shown in example (iii), below. A computer programme based on these equations is given in Table 5.1.* This programme gives ml of 1M $HA^{(n-1)-}$ and ml of 1M A^{n-} needed to make 1 litre of buffer solutions having a specified ionic strength and lying within the range $pH = pK_a' \pm 1$. If n is set equal to zero, the programme applies to BH^+/B, while $n = -1, -2$, etc., are for BH^{2+}/B^+, BH^{3+}/B^{2+}, etc. This computer programme was tested for a number of buffer systems and was found to give good agreement with published values. For example, it

---

*(All computer programmes in this Book are written in FOCAL, a 'conversational' Fortran-type language developed by Digital Equipment Corporation (Maynard, Mass., U.S.A.) for use with their small computers. FOCAL is closely related to BASIC, and anyone familiar with BASIC or Fortran will have no difficulty in translating the programmes into these languages. All Figures in this book were drawn by a Hewlett-Packard 7200A Graphic Plotter operating under computer control).

reproduced to within ±0.05 pH units the pH-composition tables of constant ionic strength buffers given by Long (1961).

5.1.2 Constant ionic strength buffers with added electrolyte.

Calculations are essentially similar for constant ionic strength buffers, initially of ionic strength I, to which an indifferent electrolyte such as NaCl is added to change the ionic strength to I'. The value of pK_a' is obtained from Equations 5.2 or 5.3 by inserting the value of I', and Equations 5.6, 5.7, 5.8 remain as before, with $I = I' - [NaCl]$.

A more common situation is one in which the sum of the concentrations of $HA^{(n-1)-}$ and A^{n-} remains constant and an electrolyte is added in varying amount so as to maintain I at some specified value such as 0.1 or 0.15. As before, pK_a' is calculated from pK_a and the total ionic strength, and hence the value of x can be obtained at the given pH. Then if c is the total concentration of $HA^{(n-1)-} + A^{n-}$

$$[HA^{(n-1)-}] = c/(1+x) \qquad (5.9)$$

$$[A^{n-}] = cx/(1+x) \qquad (5.10)$$

and the concentration of NaCl needed in the solution is

$$[NaCl] = I - 0.5c\{n^2 - n + (n^2 + n)x\}/(1+x) \qquad (5.11)$$

Table 5.2 is a computer programme which, given the thermodynamic pK_a value, the final ionic strength, the total buffer concentration and the number of negative charges on the basic species (including zero for monoamines), tabulates buffer compositions as ml of 1M $HA^{(n-1)-}$, 1M A^{n-} and 1M NaCl to give 1 litre of the desired buffer solutions covering the range pH = $pK_a' \pm 1$. Like Table 5.1, insertion of negative values of n converts the programme for use with cationic buffers.

Buffer tables are frequently expressed as the titration of the acidic buffer species with alkali, or of the basic buffer species with strong acid, followed by the addition of an indifferent electrolyte to maintain a specified constant ionic strength. Tables 5.3 and 5.4 are computer programmes which generate such tabulations for $pH = pK_a' \pm 1$, given the thermodynamic pK_a, the number of negative charges on the basic species (the number may be positive, zero or negative), the total buffer concentration in moles per litre, and the specified final ionic strength. Output gives ml of 1M NaOH or HCl, and ml 1M NaCl, at each pH for 1 litre of buffer solution.

5.1.2.1 Preparation of amine buffers of constant ionic strength.

In preparing amine buffer solutions maintained at constant ionic strength, I, by added NaCl, it is convenient to use two stock solutions. One of these contains the amine hydrochloride at a concentration A moles 1^{-1}. and sodium chloride at $(I - A)$ moles 1^{-1}. The other solution contains sodium hydroxide at a convenient concentration (say $4A - 5A$ moles 1^{-1}) and sodium chloride at I moles/1. Mixing the two solutions in any ratio (so long as the amount of sodium hydroxide does not exceed that of the amine hydrochloride) gives a final solution of constant ionic strength, I. (For a worked example, see (iv) below.)

5.1.3 Buffers by direct titration of weak bases (or acids) with strong acids (or bases).

Buffers formed by the partial neutralization of weak acids or bases vary in ionic strength. Thus, the ionic strength of an acetic acid solution to which sodium hydroxide is added is equal to the concentration of sodium hydroxide, and for buffers of organic amines with hydrochloric acid the ionic strength is equal to the concentration of added acid. This variation in ionic strength causes changes in pK_a' values with pH so that Tables 5.1 and 5.2 do not apply. To calculate the buffer composition at any pH it is necessary to use an iterative procedure, beginning with the thermodynamic pK_a and the pH to obtain a first estimate of the concentration ratio of the two buffer species, hence their individual concentrations and an estimate of the ionic strength. The latter is used to obtain an approximate 'practical' pK_a' and the cycle is repeated until adequate accuracy is obtained. Usually two cycles should be sufficient. (See Example (v), below.)

Alternatively, the desired buffer tables can be constructed rapidly using the computer programmes given in Tables 5.5 and 5.6. The first of these is for the titration of $HA^{(n-1)-}$ to A^{n-} with NaOH, the input comprising the thermodynamic pK_a, the number of negative charges (which may be positive, zero or negative) on the basic species, and the total concentration of buffer species. The programme uses Davies' equation and iterates to find pK_a', the concentration ratio of basic and acidic buffer species, and I at each pH in steps of 0.1 pH unit for $pH = pK_a' \pm 1$. The output tabulates, at each pH, the ml of 1M NaOH needed for 1 litre of buffer solution, and also I. Similarly, Table 5.6 enables the calculation of volumes of 1M hydrochloric acid needed to adjust the pH of 1 litre of buffer solution for the titration of A^{n-} to $HA^{(n-1)-}$. As an example, Table 5.6 has been used to compute the composition of 0.05M Tris buffers at 25°C for the addition of hydrochloric acid. Results in Table 5.7 are almost identical with the published values given in Table 10.32.

The pH values of acetic acid-Tris buffers have been determined experimentally (Pilz and Johann, 1966) for mixtures of 50 ml 0.2M Tris, x ml 0.2M acetic acid, diluted to a final volume of 100 ml. As examples, $x = 40$, 30 and 20 ml for pH 7.70, 8.07 and 8.42, respectively. Computed values, using Table 5.6 and taking the thermodynamic pK_a of Tris as 8.20 at 20°C, agree within ±0.05 pH units with the experimental ones. For buffers of twice this concentration there is similar agreement below pH 8.3 but at higher pH values the difference reaches 0.1 pH units.

Examples (i). What is the pH of a solution comprising a mixture of 0.0045M furoic acid and 0.000 48M sodium furoate at 25°C, given that the pK_a of furoic acid is 3.17? (This problem illustrates the use of the extended Henderson-Hasselbalch equation).

As a first approximation, pH ≈ pK_a + log (0.000 48/ 0.0045) = 2.2, giving [H⁺] ~ 0.006.

Hence I ≈ 0.006 + .000 48 = 0.0065, giving pK_a' ≈ 3.17 − 0.037 (from Table 2.3) = 3.13.

Substitution in Equation 10.15 gives [H⁺] ≈ 1.32 × 10⁻³, so that pH = 2.88.

Repeating the calculation with this new estimate gives I ≈ 0.001 32 + 0.000 48 = 0.0018, and pK_a' = 3.17 − 0.20 = 3.15.

Substitution in Equation 10.15 gives [H⁺] = 1.26 × 10⁻³, giving a final pH of 2.90.

(This agrees with the experimental value given in Table 5.8.)

(ii) What would be the pH of a buffer made by mixing 50 ml 1M KCl, 50 ml 1M boric acid, and 30 ml 1M NaOH, and diluting to 1 litre, at 25°C? (This illustrates the use of the simple Henderson-Hasselbalch equation).

In the notation of Appendix IV, C_a = 0.05 − 0.03 = 0.02, and C_b = 0.03. The value of I is approximately [KCl] + C_b = 0.08.

The pK_a of boric acid = 9.23. Hence, from Table 2.3, pK_a' = 9.23 − 0.10 = 9.13.

This gives the pH of the buffer (= pK_a' + log C_b/C_a) = 9.13 + 0.18 = 9.31. The pH thus lies within the range given in Appendix IV for which this equation is adequate. The calculated value agrees well with the experimental value given in Table 10.37.

(iii) Calculate the composition of a phosphate buffer of ionic strength 0.1 and pH 7.0 at 25°C, containing no added electrolytes.

From Appendix III, the pK_a for the loss of a proton from $H_2PO_4^-$ is 7.20 at 25°C. For an ionic strength of 0.1 the correction term to convert pK_a to pK_a' is -0.11×3 (from Table 2.3, with $n = 2$; see note to Table). Hence $pK_a' = 6.87$.

The desired pH = 7.0, so, from Equation 5.4, the ratio of $[HPO_4{}^{2-}]/[H_2PO_4{}^-]$ = antilog $(0.13) = 1.35$.

From Equation 5.6, the composition of the buffer is thus given by $[H_2PO_4{}^-] = 0.1/5.05 = 0.0198M$, and $[HPO_4{}^{2-}] = 0.135/5.05 = 0.0268M$.

(This composition is closely similar to the somewhat more dilute buffer given in Table 10.25).

(iv) Calculate the composition of a Tris buffer approximately 0.02M in total Tris + Tris · HCl, ionic strength 0.15 with NaCl, pH = 7.5 at 37°C.

From Appendix III the pK_a of Tris is 8.06 at 25°C and $d(pK_a)/dT = -0.028$, hence the thermodynamic pK_a of Tris at 37°C is $8.06 - 0.028 \times 12 = 7.72$. From Table 2.3, for an ionic strength of 0.15, $pK_a' = 7.72 + 0.12 = 7.84$, and from Table 2.1 the ratio of [Tris]/[Tris.HCl] in the mixture is 0.45 (for a pH less by 0.35 than the pK_a'), corresponding to 31% neutralization of the initial Tris.HCl (from the last column of Table 2.1).

A suitable buffer mixture would therefore be made by mixing 100 ml 0.02M Tris.HCl, 0.13M in NaCl, with 6.2 ml 0.1M NaOH, 0.15M in NaCl.

(v) What is the composition of an *N*-ethylmorpholine buffer of total concentration 0.1M and pH 7.8 at 25°C, made by adding HCl to the free base?

From Appendix III, the pK_a of *N*-ethylmorpholine is 7.67 at 25°C.

From pH = pK_a' + log [basic species]/acidic species], setting as a first approximation $pK_a \approx pK_a'$ gives $C_b/C_a \approx 10^{7.80-7.67} = 1.35$.

But $C_a + C_b = 0.1$, so $C_a \approx 0.1/2.35 = 0.043M$, and $C_b \approx 0.057M$. This gives $I = C_a \approx 0.043$.

From Table 2.3, $pK_a' \approx 7.67 + 0.08 = 7.75$.

The new ratio of C_b/C_a becomes $10^{0.05} = 1.12$.

The new values of C_a and C_b are thus $C_a = 0.0472M$, $C_b = 0.0528M$.

The new value of $I = 0.047$, and this does not lead to any further significant change in pK_a'.

Thus the desired buffer would be obtained by mixing 50 ml of an 0.2M *N*-ethylmorpholine solution with 47.2 ml of 0.1M HCl, followed by dilution to 100 ml.

5.2 On designing a new pH-buffer system

The typical pK_a values of many acidic and basic groups fall within fairly narrow pH ranges as shown in Table 5.9. By appropriate molecular modifications, these values can be varied by several pH units. Examples include the change from acetic acid (pK_a 4.76) to trifluoroacetic acid (pK_a −0.26), or from benzoic acid (pK_a 4.20) to 2,4,6-trinitrobenzoic acid (pK_a 0.65). Higher pK_a values can be obtained for the second and subsequent ionizations of polycarboxylic acids but large ionic strength effects make such compounds less suitable as buffers. Larger differences are possible with phenols. Whereas the pK_a of phenol is 10.00, for picric acid it is 0.22. These, and similar, pK_a ranges set limits to the possible types of buffers that can be designed.

In proposing the use of a substance as a buffer it is desirable that the material is commercially available, or is easily synthesized, that it is readily purified, that it is not expensive and that it is stable. It is, preferably, optically transparent to ultraviolet and visible light, it does not form strong complexes or insoluble salts with metal ions and, to diminish ionic strength effects, the buffer cation or anion is univalent.

Good *et al.* (1966) laid down some further criteria for the design of new buffers for biological research. They were concerned with pK_a values between 6 and 8, because this corresponds to the pH region where most biological reactions occur and where fewest buffers are available. Solubility was required to be high in water but low in other solvents so that in particulate systems there would be very little of the buffer inside the particulate phase, and so that the buffers would pass only with difficulty through biological membranes. High water-solubility also permits the use of aliquots of stock buffer solutions. Buffers should resist enzymatic and non-enzymatic degradation under experimental conditions and should be sufficiently different from enzyme substrates that they do not act as analogue inhibitors.

These considerations led Good et al. to examine a number of amino acids derived from substituted glycines or taurines. Some typical 'Good' buffers are MES, ADA and HEPES.

$$O \overset{+}{N}HCH_2CH_2SO_3^- \qquad H_2NCOCH_2\overset{+}{N}H \underset{CH_2COO^-Na^+}{\overset{CH_2COO^-}{<}}$$

MES (pK_a' 6.15)   ADA (pK_a' 6.6)

$$HOCH_2CH_2\overset{+}{N}H \quad NCH_2CH_2SO_3^-$$

HEPES (pK_a' 7.55)

Similarly, taurine (2-aminoethylsulphonic acid) might well be considered as an alternative to borates in buffers around pH 9. Its pK_a at 25°C is 9.06 (compare 9.23 for boric acid), and taurine is inexpensive, available pure, readily soluble and stable to drying at 110°C. Tables 10.1 and 10.3 should be applicable.

Other examples of possible zwitterionic buffers can be illustrated by PIPPS, PIPBS and EDPS (Jermyn, 1967). EDPS is also a strong complexing agent for copper. N-(2-Acet-amido)glycine, $H_2NCOCH_2\overset{+}{N}H_2CH_2COO^-$, pK_a 7.7, is a potentially useful buffer but it is not readily available.

$$HO_3SCH_2CH_2CH_2N \quad NCH_2CH_2CH_2SO_3H$$

PIPPS, 1,4-bis(3-sulphopropyl)piperazine

$pK_a' = 4.05, 8.1$ at 18°C and $c = 0.05M$

$$HO_3S(CH_2)_4N\bigcirc N(CH_2)_4SO_3H$$

PIPBS, 1,4-bis(4-sulphobutyl)piperazine

$pK_a' = 4.6, 8.6$ at $18°C$ and $c = 0.05M$

soly. 0.3 g 100 ml^{-1} at $1°C$

$$HO_3SCH_2CH_2CH_2NHCH_2CH_2NHCH_2CH_2CH_2SO_3H$$

EDPS, *N,N'*-bis(3-sulphopropyl)ethylenediamine

$pK_a' = 6.65, 9.8$ at $18°C$

Mono-tris and Bis-tris are examples of buffer substances produced by minor modification of an existing buffer, Tris (Lewis, 1966).

$H_2NC(CH_2OH)_3$	Tris pK_a ($25°C$) 8.06
$(HOCH_2CH_2)HNC(CH_2OH)_3$	Mono-tris pK_a ($25°C$) 7.83
$(HOCH_2CH_2)_2NC(CH_2OH)_3$	Bis-tris pK_a' ($20°C$) 6.46

In the related series, ethanolamine (pK_a 9.50), diethanolamine (pK_a 8.88) and triethanolamine (pK_a 7.76), ethanolamine and triethanolamine have been suggested as buffers, singly, in admixture, and as their phosphates, for the pH range 5.6 to 11.4. (See Thies and Kallinich (1953) for tables). Schwabe *et al* (1959) describe the buffer properties of other alkanolamines.

Table 5.1 *Computer programme for calculating composition of constant ionic strength buffer.*

```
Ø1.Ø1 TYPE !!!!
Ø1.Ø5 ASK ?PK?,?I?,?N?,!!!
Ø1.1Ø SET PK=PK-(2*N-1)*(Ø.5*FSQT(I)/(1+FSQT(I))-Ø.1*I)
Ø1.2Ø SET Z=FITR(1Ø*PK);SET PH=Z/1Ø-1
Ø1.3Ø TYPE "  PH    ML A    ML B",!!
Ø1.35 SET X=FEXP((PH-PK)*2.3Ø3)
Ø1.4Ø IF (N-1) 2.Ø3,3.1,3.1


Ø2.Ø3 IF (Ø-N-1) 2.1;SET A=I/(3+X)
Ø2.Ø5 SET B=X*A;GOTO 4.1
Ø2.1Ø SET A=I;SET B=X*I;GOTO 4.1


Ø3.1Ø SET A=I/(N-1+(2*N-1)*X);SET B=X*A;GOTO 4.1


Ø4.1Ø TYPE %4.Ø2,PH,"   ",1ØØØ*A," ",1ØØØ*B,!
Ø4.2Ø SET PH=PH+Ø.1;IF (PK+1-PH) 4.3,1.35,1.35
Ø4.3Ø TYPE !!!;QUIT
*
```

Table 5.2 *Computer programme for calculating composition of buffers having constant total buffer concentration and constant ionic strength.*

```
Ø1.Ø1 TYPE !!!!
Ø1.Ø5 ASK ?PK?,?I?,?N?,?C?,!!!
Ø1.1Ø SET PK=PK-(2*N-1)*(·5*FSQT(I)/(1+FSQT(I))-Ø.1*I)
Ø1.2Ø SET Z=FITR(1Ø*PK);SET PH=Z/1Ø-1
Ø1.3Ø TYPE "  PH      ML A     ML B ML NACL",!!
Ø1.35 SET X=FEXP((PH-PK)*2.3Ø3);SET A=C/(1+X);SET B=X*A
Ø1.4Ø SET IB=Ø.5*(N*(N-1)+(N↑2+N)*X)*C/(1+X)
Ø1.5Ø SET T=I-IB


Ø2.1Ø TYPE %4.Ø2,PH,"   ",1ØØØ*A,"   ",1ØØØ*B,"   "1ØØØ*T,!
Ø2.2Ø SET PH=PH+Ø.1;IF (PK+1-PH) 2.3,1.35,1.35
Ø2.3Ø TYPE !!!;QUIT
*
```

Table 5.3 *Computer programme for buffers by titration with NaOH and addition of NaCl to give constant ionic strength.*

```
Ø1.Ø5 TYPE !!!!;ASK ?PK?,?N?,?C?,?I?,!!!;SET Z=FITR(1Ø*PK);SET PH=Z/1Ø-1

Ø1.Ø7 TYPE "    PH    ML NAOH    ML NACL",!!

Ø1.1Ø SET PK=PK-(2*N-1)*(.5*FSQT(I)/(1+FSQT(I))-Ø.1*I)

Ø1.2Ø SET X=FEXP((PH-PK)*2.3Ø3);SET A=C*X/(1+X);SET B=C/(1+X)

Ø1.3Ø SET I1=Ø.5*(N*(N-1)+(N↑2+N)*X)*C/(1+X)


Ø2.2Ø TYPE %4.Ø1,PH,"    ",1ØØØ*A,"    ",1ØØØ*(I-I1),!

Ø2.3Ø SET PH=PH+Ø.1; IF (PK+1-PH) 2.4,1.2,1.2

Ø2.4Ø TYPE !!!;QUIT

*
```

Table 5.4 *Computer programme for buffers by titration with HCl and addition of NaCl to give constant ionic strength.*

```
Ø1.Ø5 TYPE !!!!;ASK ?PK?,?N?,?C?,?I?,!!!;SET Z=FITR(1Ø*PK);SET PH=Z/1Ø-1

Ø1.Ø7 TYPE "    PH    ML HCL    ML NACL",!!

Ø1.1Ø SET PK=PK-(2*N-1)*(Ø.5*FSQT(I)/(1+FSQT(I))-Ø.1*I)

Ø1.2Ø SET X=FEXP((PH-PK)*2.3Ø3);SET A=C*X/(1+X);SET B=C/(1+X)

Ø1.3Ø SET I1=Ø.5*(N*(N-1)+(N↑2+N)*X)*C/(1+X)+Ø.5*B


Ø2.2Ø TYPE %4.Ø1,PH,"    ",1ØØØ*B"    ",1ØØØ*(I-I1),!

Ø2.3Ø SET PH=PH+Ø.1;IF (PK+1-PH) 2.4,1.2,1.2

Ø2.4Ø TYPE !!!;QUIT

*
```

Table 5.5 *Computer programme for calculating composition of buffers made by direct titration of a weak acid with alkali.*

```
Ø1.Ø5 TYPE !!!!;ASK ?PK?,?N?,?C?,!!!;SET Z=FITR(1Ø*PK);SET PH=Z/1Ø-1

Ø1.Ø7 TYPE "    PH      ML NAOH      I",!!

Ø1.1Ø SET X1=FEXP((PH-PK)*2.3Ø3)

Ø1.2Ø SET A1=C*X1/(1+X1),SET B1=C/(1+X1)

Ø1.3Ø SET I1=Ø.5*(N*(N-1)+(N↑2+N)*X1)*C/(1+X1)

Ø1.4Ø SET PA=PK-(2*N-1)*(Ø.5*FSQT(I1)/(1+FSQT(I1))-Ø.1*I1)

Ø1.5Ø SET X2=FEXP((PH-PA)*2.3Ø3)

Ø1.6Ø IF (FABS((X2-X1)/X2)-1E-5) 2.1, 2.1;SET X1=X2;GOTO 1.2


Ø2.1Ø SET I=I1;SET A=A1;SET B=B1

Ø2.2Ø TYPE %4.Ø2,PH," "1ØØØ*A,"     ",%5.Ø3,I,!

Øs.3Ø SET PH=PH+Ø.1;IF (PK+1-PH) 2.4,1.1,1.1

Ø2.4Ø TYPE !!!;QUIT

*
```

Table 5.6 *Computer programme for calculating composition of buffers made by direct titration of a weak base with HCl.*

```
Ø1.Ø5 TYPE !!!!;ASK ?PK?,?N?,?C?,!!!;SET Z=FITR(1Ø*PK);SET PH=X/1Ø-1

Ø1.Ø7 TYPE "   PH     ML HCL      I",!!

Ø1.1Ø SET X1=FEXP((PH-PK)*2.3Ø3)

Ø1.2Ø SET A1=C*X1/(1+X1);SET B1=C/(1+X1)

Ø1.3Ø SET I1=Ø.5*(N*(N-1)+(N↑2+1)*X1)*C/(1+X1)+Ø.5*B1

Ø1.4Ø SET PA=PK-(2*N-1)*(Ø.5*FSQT(I1)/(1+FSQT(I1))-Ø.1*I1)

Ø1.5Ø SET X2=FEXP((PH-PA)*2.3Ø3)

Ø1.6Ø IF (FABS((X2-X1)/X2)-1E-5) 2.1,2.1;SET X1=X2;GOTO 1.2


Ø2.1Ø SET I=I1;SET A=A1;SET B=B1

Ø2.2Ø TYPE %4.Ø2,PH,"    ",1ØØØ*B,"     ",%5.Ø3,I,!

Ø2.3Ø SET PH=PH+Ø.1;IF (PK+1-PH) 2.4,1.1,1.1

Ø2.4Ø TYPE !!!;QUIT

*
```

Table 5.7 *Computed compositions of 0.05M Tris buffers at 25°C, for addition of 1M HCl to give 1 litre of final solution.*

pH	ml HCl	I	pH	ml HCl	I
7.00	46.65	0.047	8.10	25.77	0.026
7.10	45.85	0.046	8.20	22.80	0.023
7.20	44.88	0.045	8.30	19.89	0.020
7.30	43.71	0.044	8.40	17.11	0.017
7.40	42.32	0.042	8.50	14.53	0.015
7.50	40.68	0.041	8.60	12.19	0.012
7.60	38.78	0.039	8.70	10.12	0.010
7.70	36.61	0.037	8.80	8.33	0.008
7.80	34.18	0.034	8.90	6.80	0.007
7.90	31.53	0.032	9.00	5.52	0.006
8.00	28.70	0.029			

Table 5.8 *Furoic acid, sodium furoate buffer (25°C)**

(x ml 0.01M sodium furoate (1.3402 g l^{-1}), 50 − x ml 0.01M furoic acid (1.1203 g l^{-1}), diluted to 100 ml)

pH	x	pH	x	pH	x
2.90	4.8	3.40	28.6	3.80	39.9
3.00	10.4	3.50	32.0	3.90	41.6
3.10	16.0	3.60	35.2	4.00	43.3
3.20	21.6	3.70	37.7	4.10	44.6
3.30	25.2				

*German & Vogel (1937)

Table 5.9 *Typical pK$_a$ ranges for acidic and basic groups*

Acidic group	pK$_a$	Basic group	pK$_a$
α-Amino acids	2−3	Pyrimidines	1−2
Monocarboxylic acids	3−5	Anilines	3−5
Aliphatic dicarboxylic acids	2−4.5, 5−7	Aminoethers	4−5
Thiophenols	5−7	Pyridines	4−6
Hydroxy heteroaromatics	7−11	α, β-Unsaturated aliphatic amines	6−9
Phenols	8−10		
Purines	8−10	Imidazole	7
Thiols	9−11	α-Aminoacids	9−10.5
Oximes	10−12	Saturated nitrogen heterocycles	9−11
Aldehydes	12−14		
Alcohols and sugars	13−16	Aliphatic and alicyclic amines	9−11
		Guanidines	11−14

Chapter Six

Buffers for use in Partially Aqueous and Non-Aqueous Solvents and Heavy Water

Solvent molecules are involved in acid-base equilibria as acceptors or donors of protons, so that the acidic or basic strength of a substance varies with the nature of the solvent. The lower alcohols resemble water, in that they can form the ions ROH_2^+ and RO^-, but their dissociation is less than for water ($pK_{CH_3OH} = 16.7$, $pK_{C_2H_5OH} = 19.1$, cf. $pK_w = 14.0$). Consequently, substances dissolved in alcohols are weaker acids and bases than in water. Other factors influencing acidic and basic strengths in solutions include the dielectric constant and solute-solvent interactions which, in mixed solvents, can lead to the further complication of selective ordering of solvent molecules around ionic species.

Many measurements have, nevertheless, been made in mixed solvents on the basis of calibration with aqueous buffers. Thus, Simon (1964) has calculated constants (pK_{MCS}^* values) for over 1000 organic compounds in 80% (w/w) methylcellosolve/20% (w/w) water by measuring 'apparent pH values' in this solvent, based on the usual standardization of the pH meter with aqueous buffer solutions. These constants show no general correlation with the pK_a values obtained for aqueous solutions.

Similar types of measurements have been made in other organic solvents such as acetic acid, acetone, acetonitrile, cyclohexanone, dimethylformamide, dioxane, nitropyridine and pyridine (Simon, 1964). Although such pK^* values are useful for practical work, they lack thermodynamic significance. That is to say, the pH meter gives reproducible readings in other solvent systems but the pH numbers as taken from the instrument do not relate simply to the chemical equilibrium.

This has led to attempts to develop pH* scales which have a sounder theoretical basis, where the asterisk now indicates that the value is measured relative to an ideal dilute solution in the same solvent. The pH* scales are established on the basis of assigned values to particular buffers in the solvent systems, in a manner strictly analogous to the way in which the pH scale has been defined for water. Most of the work in this field has been done by Bates and his colleagues at the United States National Bureau of Standards, and by de Ligny and co-workers in the Netherlands.

6.1 pH* scales

Thus pH* scales can be set up for a wide variety of media including partially aqueous and non-aqueous solvents, soft and moist solids, and slurries. Such scales are different for each medium because of differences in solvent acidity/basicity and dielectric constant and differences in ion activities and mobilities.

From their method of measurement and definition, pH* values are consistent with the thermodynamic equations for acid-base equilibria in the solvent system to which they apply.

$$\text{For example pH*} = pK_a^* + \log \frac{\text{(basic form)}}{\text{(acidic form)}} .$$

However, pH* values for different solvents cannot be compared with one another or with pH values for aqueous solutions. Consequently, *solutions of different solvent composition may give the same pH meter reading yet behave in quite different ways in acid-base reactions.*

6.2 pH* buffers

Standard values of pH* can be assigned to buffers in partially aqueous and non-aqueous solvents but the work in establishing pH* buffer standards and pH* buffer tables to cover the working range for the many possible solvent compositions is clearly formidable. To date, most effort has been concentrated on methanol, ethanol and their aqueous mixtures. Table 6.1 gives pH* values in 50% (w/w) methanol-water at temperatures, 10–40°C, for three buffer systems developed by Paabo *et al.* (1965) as pH* standards. Similarly, Tables 6.2 and 6.3 list some selected pH* values for equimolal mixtures of Tris/Tris HCl (Woodhead *et al.*, 1965) and 4-aminopyridine/4-aminopyridine HCl (Paabo *et al.*, 1966) at temperatures, 10–40°C, in the same solvent. pH* values in various methanol-water mixtures are given in Table 6.4.

Table 6.5 comprises standard pH* values in ethanol-water mixtures. Buffers for use in absolute methanol (0.05% water) (Broser and Fleischhauer, 1970) are given in Tables 6.4 and 6.6. The latter buffers cover the pH* range, 2.4–12.4, in steps of 0.2 pH* units with a reproducibility of ±0.03 pH* units. Buffers for use in absolute ethanol are included in Table 6.5.

6.3 The measurement of pH*

Conventional pH meters can be used to measure acidity in many partially aqueous and non-aqueous media. The glass electrode responds in a reproducible way to hydrogen ions in media which contain at least a few per cent of water and also in certain anhydrous solvents. In the calomel electrode –KCl bridge system, the junction potential between the aqueous and non-aqueous medium may be large. However, this does not preclude the accurate measurement of acidity in media of constant solvent composition provided the junction potential does not vary with acidity.

pH* values in a particular medium can be determined directly by using a conventional pH meter which has been standardized using appropriate pH* standard buffers in the same solvent system. The meter reading is then in pH* units.

An approximate method of measuring pH* is also available (Bates *et al.*, 1963). If a pH meter is standardized using aqueous buffers, the meter reading, pH(R), obtained in a partially aqueous or non-aqueous medium differs by an amount, δ, from the reading, pH*, obtained when the meter is standardized using appropriate standard pH* buffers. That is, pH* = pH(R) $- \delta$. The quantity, δ, is found to be approximately constant for a given solvent concentration, temperature and solute concentration. Thus an approximate value for pH* can be obtained if values of δ are known. This method is not a procedure of high accuracy as the asymmetry potential of the glass electrode is likely to vary when the electrode is removed from the aqueous standardizing medium and then immersed in the solvent system in which measurement is to be made.

Some values of δ for partially aqueous mixtures are given in Table 6.7 (Douheret, 1967 and 1968; Reynaud, 1969; Lesquibe and Reynaud, 1970), for organic component concentrations up to limits recommended by Lesquibe and Reynaud (1970) from studies of ion-pair formation. Other approaches to the determination of δ values have been described (Ong *et al.*, 1964; Van Veen, 1971).

Comparisons have been made of the pH of aqueous buffers with pH meter readings (pH(R)) for the same concentrations of these buffers in partially aqueous solution. The differences, pH $-$ pH(R), for various concentrations of acetate/acetic acid and ammonia/ammonium buffers in water and in 50% (v/v) ethanol-water (Gottschalk, 1959) are fairly constant at 0.89–0.92 and 0.24–0.27, respectively, when the pH meter is standardized against aqueous buffers. These differences are not δ values as defined above, and tables of this kind have not been included in this chapter.

6.4 A universal pH scale

Problems involved in the establishment of a universal pH scale applicable to all solvent systems have been discussed by Bates (1964). The essential difficulty is in the independent determination of the activities of individual ions and of liquid junction potentials. More recently, de Ligny & Alfenaar

(1967) have proposed a universal pH scale, and, in a series of papers, de Ligny & co-workers have reported the activity co-efficients of single ions in a number of solvent systems (Bax *et al.*, 1973).

6.5 The pD scale and the measurement of pD

The increasing use of heavy water (D_2O) in studies of reaction mechanism, in nuclear technology and in medical research has created a demand for accurate measurements of acidity in this medium. As the glass electrode responds satisfactorily to deuterium ion in D_2O, this need is provided by modern pH meters. The operational definition of pH also allows the creation of a scale, the pD scale, where the solvent is heavy water. Reference values have been assigned to three selected buffers in heavy water (Paabo & Bates, 1969). Standard pD values for these buffers for various temperatures are given in Table 6.8. When pH meters are standardized against these buffers in heavy water solution, the meter readings are in pD units.

Exchange of ionizable hydrogen with deuterium occurs when the buffer salts are dissolved in heavy water and there is therefore a contamination of the solvent with hydrogen. The change in pD due to this change in solvent composition is, however, negligible. (For heavy water of isotopic purity of at least 99.5%, the change in pD is not more than 0.001 pD unit.)

Alternatively, if the pH meter is standardized against aqueous pH buffers, there is a simple relationship (Glasoe and Long, 1960) between pD and the meter reading:

$$pD = \text{meter reading} + 0.40.$$

These authors showed that the relation is independent of whether the internal solution of the calomel electrode contains water or heavy water. This finding also applies to the salt bridge (Covington *et al.*, 1968).

6.6 The use of pH* and pD buffers
6.6.1 The determination of dissociation constants of acids.

Some acids and bases are of such low solubility in water that the concentrations attainable are insufficient for potentiometric or spectrophotometric determination of dissociation constants. This difficulty may be overcome by the use of partially aqueous solvent mixtures and non-aqueous solvents in which the compound is more soluble. In a large proportion of cases, subsequent calculations of dissociation constants have been based on the readings of pH meters standardized against aqueous pH buffers. Such constants should be reported as 'apparent dissociation constants' for the particular medium, $pK_a(app)$. Simon's (1964) pK^*_{MCS} values belong to this category.

If measurements are made in a mixed solvent system using a pH meter standardized with appropriate pH* buffers, pH* readings are obtained and a thermodynamic dissociation constant for that medium (pK_a*) can be calculated. An approximate value of pK_a* can also be obtained if the appropriate δ value is known.

Having obtained a series of pK_a* values for varying solvent mixtures, it should be possible to extrapolate to the pK_a value of the sparingly soluble acid or base in water. This approach appears preferable to the plot of pH meter readings, pH(R), for the mixed-solvent system against $pK_a(app)$, followed by extrapolation to pure water.

For further discussion of dissociation constant measurement in mixed and non-aqueous solvents, see Albert and Serjeant (1971) and Van Veen *et al.*, (1971).

For pK_a determinations in heavy water, pD values are obtained using a pH meter calibrated either against standard pD buffers, or against pH buffers and then applying the correction: pD = pH + 0.4.

6.6.2 Rate studies in heavy water.

The alkaline hydrolysis of esters in heavy water can be studied by NMR spectroscopy (Fenn, 1973). The pseudo first order rate constant for the reaction includes the term, $[OD^-]$. The pD can be determined as in the previous paragraph. From pD, $[OD^-]$ can be calculated using the autoprotolysis constant for heavy water, $K_{D_2O} = 14.9$ (Covington *et al.*, 1966), and hence the true rate constant is obtained.

6.7 Surfactants

Aqueous solutions containing appreciable quantities of surfactants can be looked on as special cases of partially aqueous systems. Making a solution 3% (w/v) in cetomacrogol (a nonionic detergent), 3% (w/v) in sodium dodecyl sulphate (an anionic surfactant) or 0.3% (w/v) in cetyl pyridinium bromide (a cationic surfactant) did not significantly change the pH of phthalate, phosphate or borate buffers as measured by a glass electrode. The maximum effect, a decrease of 0.14 pH unit, was found for cetyl pyridinium bromide with the phthalate buffer (Florence and Dempsey, 1973). At the ionic strengths involved, these surfactants were above their critical micelle concentration.

Table 6.1 *Standard pH* values* in 50% (w/w) methanol-water†*

	HOAc, NaOAc, NaCl Each component		NaHSuc‡, NaCl Each component		KH_2PO_4, Na_2HPO_4, NaCl Each component	
$T°C$	0.01m pH*	0.05m pH*	0.01m pH*	0.05m pH*	0.01m pH*	0.02m pH*
10	5.586	5.518	5.863	5.720	8.072	7.937
15	5.577	5.506	5.844	5.697	8.051	7.916
20	5.571	5.498	5.829	5.680	8.034	7.898
25	5.568	5.493	5.818	5.666	8.021	7.884
30	5.569	5.493	5.811	5.656	8.011	7.872
35	5.573	5.496	5.806	5.650	8.004	8.863
40	5.580	5.502	5.806	5.648	8.001	7.858

Estimated uncertainty, ±0.01 pH units
†Paabo *et al.* (1965)
‡succinate

Table 6.2 *Selected*[†] *pH** *values*[‡] *for equimolal Tris-Tris hydrochloride in 50% (w/w) methanol-water*[§]

	Molality			
T°C	0.01 pH*	0.02 pH*	0.05 pH*	0.10 pH*
10	8.35	8.38	8.44	8.49
15	8.19	8.22	8.28	8.33
20	8.04	8.07	8.13	8.18
25	7.90	7.93	7.99	8.04
30	7.76	7.79	7.85	7.90
35	7.63	7.66	7.72	7.77
40	7.51	7.54	7.60	7.65

[†] Reference gives values from 0.01m to 0.10m at 0.01m intervals
[‡] Estimated uncertainty, ±0.05 pH* units.
[§] Woodhead, *et al.* (1965).

Table 6.3 *pH* values for equimolal 4-aminopyridine/4-aminopyridine hydrochloride in 50% (w/w) methanol-water*[†]

	Molality				
T°C	0.02 pH*	0.04 pH*	0.06 pH*	0.08 pH*	0.10 pH*
10	9.048	9.088	9.116	9.137	9.155
15	8.901	8.941	8.968	8.989	9.007
20	8.762	8.802	8.829	8.850	8.866
25	8.629	8.668	8.695	8.715	8.732
30	8.500	8.541	8.570	8.596	8.608
35	8.376	8.418	8.446	8.467	8.484
40	8.262	8.305	8.332	8.353	8.368

[†] Paabo *et al.* (1966)

Table 6.4 *pH* Buffers for methanol-water mixtures (25°C)*

% (w/w) Methanol	pH*	Buffer components
0	1.76†	0.019 92m HCl
	2.05†	0.010 00m KCl, 0.010 02m HCl
	2.62†	0.017 62m KCl, 0.010 00m H$_3$Cit
	3.83†	0.009 995m KCl, 0.019 99m H$_2$Suc, 0.009 995m NaHSuc
	4.69†	0.0100m KCl, 0.010 67m HOAc, 0.010 67m NaOAc
	5.53†	0.012 00m KCl, 0.001 997m NaHSuc, 0.001 997m Na$_2$Suc
	7.01†	0.010 00m KCl, 0.002 496m KH$_2$PO$_4$, 0.002 496m Na$_2$HPO$_4$
	8.14†	0.002 000m Tris, 0.002 000m Tris-hydrochloride
	9.17†	0.009 992m KCl, 0.004 996m Na borate
	10.08†	0.0100m KCl, 0.002 514m NaHCO$_3$, 0.002 514m Na$_2$CO$_3$
8.1	2.05†	0.0100m KCl, 0.010 15m HCl
	3.96†	0.009 995m KCl, 0.019 99m H$_2$Suc, 0.009 995m NaHSuc
	4.80†	0.0100m KCl, 0.010 904m HOAc, 0.010 904m NaOAc
	5.67†	0.012 00m KCl, 0.002 058m NaHSuc, 0.002 058m Na$_2$Suc
	7.19†	0.010 00m KCl, 0.002 541m KH$_2$PO$_4$, 0.002 541m Na$_2$HPO$_4$
	8.11†	0.002 000m Tris, 0.002 000m Tris-hydrochloride
	9.19†	0.009 92m KCl, 0.004 996m Na borate
10	2.19‡ §	0.01m H$_2$Ox, 0.01m NH$_4$HOx
	4.30‡ §	0.01m H$_2$Suc, 0.01m LiHSuc
	9.24‖	0.01m Na borate
16.3	2.04†	0.01 000m KCl, 0.010 32m HCl
	4.11†	0.009 995m KCl, 0.019 99m H$_2$Suc. 0.009 995m NaHSuc

Table 6.4 *(cont.)*

% (w/w) Methanol	pH*	Buffer components
	4.94†	0.010 00m KCl, 0.011 00m HOAc, 0.011 00m NaOAc
	5.84†	0.012 00m KCl, 0.002 063m NaHSuc, 0.002 063m Na$_2$Suc
	7.38†	0.010 00m KCl, 0.002 578m KH$_2$PO$_4$, 0.002 578m Na$_2$HPO$_4$
	8.09†	0.002 000m Tris, 0.002 000m Tris-hydrochloride
	9.23†	0.009 92m KCl, 0.004 996m Na borate
	10.58†	0.010 00m KCl, 0.002 590m NaHCO$_3$, 0.002 590m Na$_2$CO$_3$
20	2.25‡ §	0.01m H$_2$Ox, 0.01m NH$_4$HOx
	4.48‡ §	0.01m H$_2$Suc, 0.01m LiHSuc
	9.30‖	0.01m Na borate
30	2.30‡ §	0.01m H$_2$Ox, 0.01m NH$_4$HOx
	4.67‡ §	0.01m H$_2$Suc, 0.01m LiHSuc
33.3	1.76†	0.021 13m HCl
	2.06†	0.010 00m KCl, 0.010 59m HCl
	2.90†	0.017 62m KCl, 0.010 00m H$_3$Cit
	4.43†	0.009 995m KCl, 0.019 99m H$_2$Suc, 0.009 995m NaHSuc
	5.24†	0.010 00m KCl, 0.011 20m HOAc, 0.011 20m NaOAc
	6.21†	0.012 00m KCl, 0.002 123m NaHSuc, 0.002 123m Na$_2$Suc
	7.77†	0.010 00m KCl, 0.002 642m KH$_2$PO$_4$, 0.002 642m Na$_2$HPO$_4$
	8.01†	0.002 000m Tris, 0.002 000m Tris-hydrochloride
	9.32†	0.009 992m KCl, 0.004 996m Na borate
40	2.38‡ §	0.01m H$_2$Ox, 0.01m NH$_4$HOx
	4.87‡ §	0.01m H$_2$Suc, 0.01m LiHSuc
	9.43‖	0.01m Na borate
43.3	5.06¶	0.01m KH phthalate

Table 6.4 (*cont.*)

% (w/w) Methanol	pH*	Buffer components
50	2.47‡§	0.01m H_2Ox, 0.01m NH_4HOx
	5.07‡§	0.01m H_2Suc, 0.01m LiHSuc
52.1	2.07†	0.010 00m KCl, 0.011 01m HCl
	4.81†	0.009 995m KCl, 0.019 99m H_2Suc, 0.009 995m NaHSuc
	5.64†	0.010 00m KCl, 0.011 60m HOAc, 0.011 60m NaOAc
	6.69†	0.012 00m KCl, 0.002 198m NaHSuc, 0.002 198m Na_2Suc
	7.95†	0.002 000m Tris, 0.002 000m Tris-hydrochloride
	8.21†	0.010 00m KCl, 0.002 752m KH_2PO_4, 0.002 757m Na_2HPO_4
	9.43†	0.009 92m KCl, 0.004 996m Na borate
	11.32†	0.010 00m KCl, 0.002 771m $NaHCO_3$, 0.002 771m Na_2CO_3
60	2.58‡§	0.01m H_2Ox, 0.01m NH_4HOx
	5.30‡§	0.01m H_2Suc, 0.01m LiHSuc
65	2.66‡§	0.01m H_2Ox, 0.01m NH_4HOx
	5.41‡§	0.01m H_2Suc, 0.01m LiHSuc
68.1	1.78†	0.022 83m HCl
	2.07†	0.010 00m KCl, 0.011 56m HCl
	3.24†	0.017 62m KCl, 0.010 00m H_3Cit
	5.21†	0.009 995m KCl, 0.019 99m H_2Suc, 0.009 995m NaHSuc
	·6.06†	0.010 00m KCl, 0.012 11m HOAc, 0.012 11m NaOAc
	7.21†	0.012 00m KCl, 0.002 301m NaHSuc, 0.002 301m Na_2Suc
	7.99†	0.002 000m Tris, 0.002 000m Tris-hydrochloride
	8.71†	0.010 00m KCl, 0.002 869m KH_2PO_4, 0.002 869m Na_2HPO_4
	9.52†	0.009 992m KCl, 0.004 996m Na borate

Table 6.4 (*cont.*)

% (w/w) Methanol	pH*	Buffer components
	11.41†	0.010 00m KCl, 0.002 500m NaHCO$_3$, 0.002 500m Na$_2$CO$_3$
70	2.76‡ § 5.57‡ §	0.01m H$_2$Ox, 0.01m NH$_4$HOx 0.01m H$_2$Suc, 0.01m LiHSuc
80	3.13‡ § 6.01‡ §	0.01m H$_2$Ox, 0.01m NH$_4$HOx 0.01m H$_2$Suc, 0.01m LiHSuc
90	3.73‡ § 6.73‡ §	0.01m H$_2$Ox, 0.01m NH$_4$HOx 0.01m H$_2$Suc, 0.01m LiHSuc
95	4.23‡ § 7.26‡ §	0.01m H$_2$Ox, 0.01m NH$_4$HOx 0.01m H$_2$Suc, 0.01m LiHSuc
100	5.79 ± 0.05‡ § 7.53 ± 0.05 § 8.75 ± 0.05‡ §	0.01m H$_2$Ox, 0.01m NH$_4$HOx 0.01m HSal, 0.01m NaSal 0.01m H$_2$Suc, 0.01m LiHSuc

† Bates *et al.*, (1963)
‡ de Ligny *et al.*, (1960)
§ Alfenaar & de Ligny, (1967)
‖ Popa *et al.*, (1965)
¶ Popa *et al.*, (1965a)
Ox = oxalate; Cit = citrate; Suc = succinate; Sal = salicylate

Table 6.5 *pH* Buffers for ethanol-water mixtures (25°C)*

Ethanol % (w/w)	pH*	Buffer components
0	2.146 ± 0.004†	0.01m H_2Ox, 0.01m, LiHOx
	2.75‡	0.008 24m NaCl, 0.002 006m HCl
	4.113 ± 0.004†	0.01m H_2Suc, 0.01m LiHSuc
	4.41‡	0.005 03m NaCl, 0.0201m HOAc, 0.010 05m NaOAc
	7.50‡	0.002 00m TEA, 0.004 01m TEA-hydrochloride
16.2	2.74‡	0.008 232m NaCl, 0.002 058m HCl
	4.64‡	0.005 15m NaCl, 0.0206m HOAc, 0.0103m NaOAc
	7.37‡	0.002 06m TEA, 0.004 12m TEA-hydrochloride
30	2.322 ± 0.005†	0.01m H_2Ox, 0.01m LiHOx
	4.692 ± 0.005†	0.01m H_2Suc, 0.01m LiHSuc
33.2	2.77‡	0.008 464m NaCl, 0.002 116m HCl
	4.99‡	0.0053m NaCl, 0.0212m HOAc, 0.0106m NaOAc
	7.18‡	0.002 14m TEA, 0.004 23m TEA-hydrochloride
50	2.503 ± 0.008†	0.01m H_2Ox, 0.01m LiHOx
	5.064 ± 0.008†	0.01m H_2Suc, 0.01m LiHSuc
52.0	2.79‡	0.008 840m NaCl, 0.002 210m HCl
	5.40‡	0.005 53m NaCl, 0.0221m HOAc, 0.011 05m NaOAc
	6.92‡	0.002 21m TEA, 0.004 42m TEA-hydrochloride
71.9	2.97 ± 0.01†	0.01m H_2Ox, 0.01m LiHOx
	5.70 ± 0.02†	0.01m H_2Suc, 0.01m LiHSuc
73.4	2.88‡	0.009 352m NaCl, 0.002 338m HCl
	6.06‡	0.005 85m NaCl, 0.0234m HOAc, 0.0117m NaOAc
	6.75‡	0.002 34m TEA, 0.004 68m TEA-hydrochloride
85.4	3.01‡	0.009 688m NaCl, 0.002 422m HCl
	6.65‡	0.006 05m NaCl, 0.0242m HOAc, 0.0121m NaOAc
	6.74‡	0.002 42m TEA, 0.004 84m TEA-hydrochloride

Table 6.5 (*Cont.*)

Ethanol % (w/w)	pH*	Buffer components
100	5.08[‡]	0.010 192m NaCl, 0.002 548m HCl
	8.31 ± 0.07[†]	0.01m HSal, 0.01m LiSal
	9.05[‡]	0.002 55m TEA, 0.005 10m TEA-hydrochloride
	9.95[‡]	0.006 38m NaCl, 0.0255m HOAc, 0.012 75m NaOAc
	13.2 ± 0.1[†]	0.01m Diethylbarbituric acid, 0.01m Li diethylbarbiturate

[†] Gelsema *et al.* (1965)
[‡] Bates *et al.* (1963)
Ox = oxalate; Suc = succinate; TEA = triethanolamine; Sal = salicylate.

Table 6.6 *pH* Buffers for use in methanol (0.05% water) (25°C)*[†]

100 ml of buffer of required pH* prepared using two of the solutions, A to K:

    A = 0.02M 4-toluenesulphonic acid
    B = 0.02M sodium methylate
    C = 0.02M sodium trifluoracetate
    D = 0.02M oxalic acid
    E = 0.02M lithium methylate
    F = 0.02M tribenzylamine
    G = 0.02M sodium hydrogen phthalate
    H = 0.02M salicylic acid
    I = 0.02M phenylacetic acid
    J = 0.02M lithium hydrogen succinate
    K = 0.02M Tris-maleate

	A + B			A + C			D + E	
pH*	A ml	B ml	pH*	C ml	A ml	pH*	D ml	E ml
2.4	88.9	to 100	3.8	51.0	to 100	4.6	92.3	to 100
2.6	74.1	to 100	4.0	54.6	to 100	4.8	90.1	to 100
2.8	64.5	to 100	4.2	60.2	to 100	5.0	85.5	to 100
3.0	58.8	to 100	4.4	67.8	to 100	5.2	80.0	to 100
3.2	55.7	to 100	4.6	74.1	to 100	5.4	74.9	to 100
3.4	53.8	to 100	4.8	81.0	to 100	5.6	69.9	to 100
3.6	52.9	to 100	5.0	86.6	to 100	5.8	65.4	to 100
3.8	52.4	to 100	5.2	90.9	to 100	6.0	61.9	to 100

Table 6.6   (*Cont.*)

	A + F			A + G			E + H	
pH*	F ml	A ml	pH*	G ml	A ml	pH*	H ml	E ml
5.8	52.6 to 100		6.2	51.7 to 100		6.4	95.2 to 100	
6.0	55.0 to 100		6.4	53.0 to 100		6.6	91.7 to 100	
6.2	59.5 to 100		6.6	54.8 to 100		6.8	87.0 to 100	
6.4	64.9 to 100		6.8	57.6 to 100		7.0	81.6 to 100	
6.6	71.2 to 100		7.0	61.0 to 100		7.2	75.5 to 100	
6.8	77.2 to 100		7.2	65.6 to 100		7.4	70.4 to 100	
7.0	83.7 to 100		7.4	72.7 to 100		7.6	65.4 to 100	
7.2	88.9 to 100		7.6	80.0 to 100		7.8	61.6 to 100	
7.4	92.2 to 100		7.8	85.1 to 100		8.0	58.3 to 100	
						8.2	55.9 to 100	
						8.4	54.2 to 100	
						8.6	53.1 to 100	

	B + I			E + J			E + K	
pH*	I ml	B ml	pH*	J ml	E ml	pH*	K ml	E ml
7.8	93.9 to 100		9.7	95.2 to 100		9.5	93.5 to 100	
8.0	90.5 to 100		9.8	90.9 to 100		9.6	91.7 to 100	
8.2	86.2 to 100		10.0	83.3 to 100		9.8	87.0 to 100	
8.4	81.0 to 100		10.2	76.9 to 100		10.0	81.3 to 100	
8.6	74.9 to 100		10.4	71.4 to 100		10.2	76.1 to 100	
8.8	69.2 to 100		10.6	66.7 to 100		10.4	70.2 to 100	
9.0	63.5 to 100		10.8	62.9 to 100		10.6	64.5 to 100	
9.2	59.6 to 100		11.0	60.1 to 100		10.8	59.4 to 100	
9.4	56.3 to 100		11.2	58.0 to 100		11.0	55.6 to 100	
9.6	54.1 to 100		11.4	56.8 to 100		11.2	52.6 to 100	

	B + G			B + G	
pH*	G ml	B ml	pH*	G ml	B ml
10.0	93.0 to 100		11.4	59.9 to 100	
10.2	89.7 to 100		11.6	56.8 to 100	
10.4	85.1 to 100		11.8	54.4 to 100	
10.6	79.7 to 100		12.0	52.6 to 100	
10.8	74.1 to 100		12.2	51.3 to 100	
11.0	69.0 to 100		12.3	50.8 to 100	
11.2	64.5 to 100		12.4	50.4 to 100	

†Broser & Fleischhauer (1970)

Table 6.7 *Values of δ for some partially aqueous mixtures (25°C)*

Non-aqueous component								
Acetone*	% (w/w)	17.8	36.4	45.9	55.8	65.4		
	δ	−0.04	−0.06	−0.09	−0.15	−0.63		
Acrylonitrile†	% (w/w)	17.6	35.6	44.8	54.9			
	δ	−0.11	−0.21	−0.26	−0.42			
Dimethylformamide†	% (w/w)	19.9	39.8	49.5	59.7	69.8	79.6	
	δ	+0.13	+0.33	+0.45	+0.54	+0.57	+0.57	
Dimethylsulphoxide†	% (w/w)	22.5	43.6	53.7	63.4	72.8	81.9	
	δ	+0.09	+0.34	+0.49	+0.65	+0.81	+0.83	
Dioxane*	% (w/w)	21.4	41.8					
	δ	−0.05	−0.13					
Ethanol*	% (w/w)	16.6	34.2	43.4	52.5	62.3	72.3	
	δ	0.02	0.10	0.15	0.22	0.25	0.27	
Methanol*	% (w/w)	17.4	36.1	45.8	55.8	65.9		
	δ	0.01	0.07	0.11	0.14	+0.15		
Methoxyethanol§	% (v/v)	10	20	30	40	50	60	70
	δ	−0.04	−0.06	−0.06	−0.06	−0.07	−0.10	−0.22
N-Methylpyrrolidone‡	% (v/v)	10	20	30	40	50	60	70
	δ	0.10	0.20	0.33	0.47	0.64	0.79	0.96
Tetramethylenesulphone‡	% (w/v)	14	28	42	56	70		
	δ	−0.07	−0.14	−0.20	−0.29	−0.41		

* Douheret (1967)
† Douheret (1968)
‡ Lesquibe & Reynaud (1970)
§ Reynaud (1969)

Table 6.8 *Standard pD values in heavy water**

$T^{\circ}C$	KD$_2$ Citrate† 0.05m	KD$_2$PO$_4$ + Na$_2$DPO$_4$† 0.025m, each	NaDCO$_3$ + Na$_2$CO$_3$† 0.025m, each
5	4.378	7.539	10.998
10	4.352	7.504	10.924
15	4.329	7.475	10.855
20	4.310	7.449	10.793
25	4.293	7.428	10.736
30	4.279	7.411	10.685
35	4.268	7.397	10.638
40	4.260	7.387	10.597
45	4.253	7.381	10.560
50	4.250	7.377	10.527

*Paabo & Bates (1969)
†Prepared by dissolving the corresponding hydrogen compound in heavy water.

Chapter Seven

Metal-Ion Buffers

Chemical and biological reactions often depend critically on the presence of low concentrations of certain free metal ions. Concentrations less than 10^{-5} M are difficult to maintain in the presence of adventitious complexing agents, hydrolytic equilibria, adsorption and, possibly, contamination. Many problems can be overcome by the use of metal-ion buffers which provide a controlled source of free metal ions in a manner similar to the regulation of hydrogen ion concentration by pH buffers.

7.1 The concept of pM

In a solution containing a metal ion, M, and a chelating agent or ligand, L, metal complexes are formed

$$M + nL \rightleftharpoons ML_n$$

and the equilibrium can be expressed quantitatively by the appropriate stability constant

$$\beta_n = \frac{(ML_n)}{(M)(L)^n} \tag{7.1}$$

This can be written as

$$pM = \log \beta_n + \log (L)^n/(ML_n) \tag{7.2}$$

where pM is the negative logarithm of the free metal ion activity and is analogous to pH.

In Equations 7.1 and 7.2 concentrations can be used instead of activities, provided the ionic strengths of solutions do not vary too widely, so long as β_n is replaced by the corresponding concentration constant.

When the ligand is present in excess, and almost all of the metal ion is bound in a complex, the free metal ion is 'buffered' in a manner similar to hydrogen ions in pH buffers.

This similarity between pM and pH buffers holds more closely for systems containing 1:1 metal complexes because, in the presence of excess ligand, and at constant pH, the ratio of [L]/[ML] is not significantly changed on dilution and hence, from Equation 7.2, pM also remains essentially constant.

Very frequently the ligand, L, can undergo protonation,

$$H^+ + L \rightleftharpoons HL, \text{etc.,}$$

to form species that have little or no complexing ability. Because of such reactions, protons compete with metal ions for the ligand. The extent to which a metal complex is formed depends in most cases on the pH of the solution and decreases as the solution is made more acidic. It is important to note that the apparent stability of metal complexes at any specified pH depends not only on the stability constant of the complex but also on the pK_a values of the ligand. The use of stability constants alone in comparing complexes of metal ions with different ligands need not give their relative orders of stability at, say, pH 7. In general it is desirable for metal-ion buffers to be pH-buffered as well, and care must be taken that the pH-buffer species is not, itself, a strong chelating agent.

7.2 Uses of metal-ion buffers

One of the major applications of pM buffers is in maintaining concentrations of necessary metal ions in biological nutrient media at essentially constant levels. As free metal ions are removed from the system, perhaps by hydrolysis or by incorporation into metalloenzymes, they are replenished by the reversible dissociation from a reservoir of metal complex. Among the first complexing agents used in this way were citrate and tartrate ions, but more recently aminopoly-carboxylic acids such as ethylenediaminetetraacetic acid (EDTA), diethylenetriaminepentaacetic acid (DTPA) and nitrilotriacetic acid (NTA) have become the chelating agents of choice. Metal-ion buffers are useful in regulating enzyme synthesis and activity *in vivo*, in studies of enzyme activation and inhibition, in obtaining information on the nature of metal-binding groups in enzymes and coenzymes, and in examining reaction rates and stabilities of many enzymes.

This metal-buffering is important when hydrolysis or insolubility limits the attainable free metal ion concentration. Thus, in toxicity studies of lead ion, the presence of phosphate and chloride ions in the physiological media and the very small solubility products of $Pb_3(PO_4)_2$ (10^{-42}), $PbHPO_4$ ($10^{-9.9}$) and $PbCl_2$ severely limit the possible concentration of free lead ion. It is not feasible to add directly to the medium a soluble lead salt, such as the nitrate, in sufficiently small amounts that precipitation can be avoided but it is practicable to calculate the lowest value of pPb that would not result in precipitation and to use chelating agents to maintain this level in the system.

Maintenance of constant pM levels may also be important in controlling the catalytic action of metal ions in industrial chemical processes. A distinction is made between metal-ion buffering and the masking of metal ions. The latter topic, discussed elsewhere (Perrin, 1970), is concerned with the use of chelating agents to diminish metal ion concentrations below levels at which they exert specified chemical or biological effects.

A promising application of metal-ion buffers is in the standardization of ion-selective electrodes.

7.3 Calculation of pM

Calculation of pM for metal-ion buffers containing excess of complexing agent is straightforward (Ringbom, 1963), using Schwarzenbach's (1957) α-coefficient method. It is necessary to know the pH of the solution, the pK_a values of the ligand and the stability constants of the metal complexes. If the metal ion undergoes significant hydrolysis the appropriate constants are also included.

'Apparent' or 'conditional' stability constants, β_n', are defined in terms of the equilibrium between a metal complex and its components, except that free ligand concentration is replaced by total concentration of all ligand species not actually complexed to the metal, and the free metal ion term includes hydrolysed metal ion and metal ion bound to other complexing species. The relation is

$$\beta_n' = \frac{\beta_n}{\alpha_M \cdot (\alpha_L)^n} \tag{7.3}$$

where $\alpha_M = ([M] + [MOH] + [M(OH)_2] + \ldots)/[M]$ and
$\alpha_L = ([L] + [HL] + [H_2L] + \ldots)/[L]$.

Many metal ions hydrolyse to form polynuclear species, so that α_M would be concentration-dependent, but in the presence of excess strong ligand it is usually sufficient to consider only the formation of mononuclear species. Under these conditions α_M reduces to

$$\alpha_M = 1 + 10^{(pH - pK_1)} + 10^{(2pH - pK_1 - pK_2)} + \ldots \tag{7.4}$$

where pK_1, pK_2, .. are the successive pK_a values for the loss of a proton from a hydrated metal ion. Some useful pK_a values are given in Table 7.1. Similarly,

$$\alpha_L = 1 + 10^{(pK_1 - pH)} + 10^{(pK_1 + pK_2 - 2pH)} + \ldots \tag{7.5}$$

where the pK_a values are for the successive loss of protons from HL, H_2L, etc. Values for some of the common ligands are given in Table 7.2. Stability constants for some of the more important metal-complexes with these ligands are listed in Table 7.3. Extensive compilations are available (Sillén and Martell, 1964, 1971).

The α-coefficient method is very convenient for calculating free metal ion concentrations when excess of a complexing agent is present. Knowing α_L and α_M from Equations 7.5 and 7.4 enables β_n' to be obtained from Equation 7.3. For a total metal ion concentration $[M]_T$, if most of the metal ion is present as a complex, ML_n, and if the total concentration of ligand species not complexed to the metal ion is $[L]_T$, we have, from the definition of β_n'

$$\beta_n' \approx [M]_T/(\alpha_M [M] [L]_T{}^n)$$

This rearranges to give

$$pM \approx \log \beta_n' + n \log [L]_T + \log \alpha_M - \log[M]_T \qquad (7.6)$$

The pH-dependence of pM is illustrated in Fig. 7.1 for a solution containing zinc ions and EDTA.

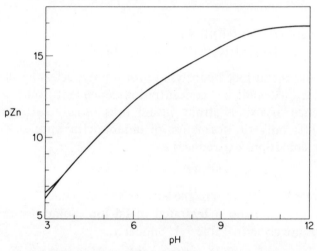

Fig. 7.1 Computed dependence of pZn on pH for a 2:1 EDTA:Zn buffer

Upper curve at pH 3 is for $[Zn]_T = 10^{-6}$ M. Lower curve is for $[Zn]_T = 0.1$M. Computation takes account of 4 pK_a values of EDTA, the hydrolysis constant of Zn^{2+}, and the stability constant of ZnEDTA.

Examples (i) What is the free calcium ion concentration in a solution 10^{-3} M in EDTA, buffered to pH 7.3 and containing 10^{-4} M calcium chloride?

From a cursory examination of Table 7.3, most of the calcium would be present as the 1:1 EDTA complex. Hence,

$$[CaEDTA] \approx 10^{-4}M, \quad [L]_T \approx 9 \times 10^{-4}M \approx (\text{total EDTA} - \text{total Ca})$$

Taking pK_a values from Table 7.2:

$$\alpha_L = 1 + 10^{(10.27-7.3)} + 10^{(16.4-14.6)} \approx 10^{3.0}$$

$$\alpha_M \approx 1$$

From Table 7.3:

$$\log \beta_1 = 10.6$$

From Equation 7.3:

$\beta_1' = \beta_1/\alpha_L = \beta_1/10^{3.0} = 10^{7.6}$

$[Ca^{2+}] = [CaEDTA]/[L]_T.10^{7.6}$ (from the definition of β_1')

$\qquad = 10^{-8.55}$

$\quad pCa = 8.55.$

(ii) What is the effect on pPb of changing the pH of a solution from 5.0 to 8.0 if it is 0.01M in NTA and 10^{-4} M in lead nitrate?

$$\alpha_L = 1 + 10^{(9.75-5.0)} = 10^{4.75} \text{ at pH 5}$$
$$= 1 + 10^{(9.75-8.0)} = 10^{1.76} \text{ at pH 8}$$
$$\log \beta_1 = 11.5$$
$$\therefore [PbNTA]/[Pb^{2+}] [L]_T = 10^{6.75} \text{ at pH 5}$$
$$= 10^{9.74} \text{ at pH 8}$$
$$[Pb^{2+}] = 10^{-8.8} \text{ at pH 5}$$
$$= 10^{-11.7} \text{ at pH 8}$$

Hence pPb increases from 8.8 to 11.7 as the pH is raised from 5 to 8.

The concentrations of hydrolysed lead species are also readily calculated from $[Pb^{2+}] \times (\alpha_M - 1)$ to be $10^{-10.9}$ M at pH 5 and $10^{-10.6}$ M at pH 8.

7.4 pH-Independent metal ion buffers

Because of the effect of the pK_a values of the ligand, in the buffers described above the free metal ion concentration varies with the pH. This pH-dependence can be overcome very simply by using a system in which there is a swamping excess of a less strongly complexing metal ion, M^{II}, so that

$$[M^{II}]_T > [L]_T > [M^I]_T.$$

Although the treatment can be made quite general, only 1:1 complexes are considered here. Let β^I be the stability constant for M^I with the ligand, and β^{II} be likewise for M^{II}. From the definitions of β^I and β^{II}, and the approximation,

$$[M^{II}] \approx [M^{II}]_T - [L]_T \tag{7.7}$$

it can be deduced that

$$[M^I] \approx \frac{[M^I]_T \beta^{II}([M^{II}]_T - [L]_T)}{\beta^I([L]_T - [M^I]_T)} \tag{7.8}$$

The expression does not contain any pH-dependent quantities and requires only that most of ligand is present as a metal complex. However, it follows from Equation 7.8 that the concentration of free metal ion M^I varies directly with the total concentration of M^I and the concentration of free metal ion M^{II}, so that pM^I varies with the dilution. The other controlling factor is the ratio of the stability constants of complexes of M^I and M^{II} with the ligand: this is a constant, the value of which depends on the particular complexing agent that is chosen.

Constancy of pM with change in pH simplifies kinetic studies, especially of enzyme systems. A further advantage of this type of buffer is that in physiological media the concentrations of such essential metal ions as calcium and magnesium can be maintained at desired levels while, by appropriate choice of chelating agent, the concentrations of other metal ions such as copper and zinc can be varied in a controlled manner. As shown in Table 7.4, the presence of excess magnesium ion enables pCa to be set at any value between 4 and 7, using EDTA. Replacement of EDTA by EGTA (ethylene glycol bis(2-aminoethylether)tetraacetic acid) would raise the listed pCa values by 3.9. The greater range of pZn values in Table 7.4 is due, directly, to the range of other metal ions in this system. It could be further extended by appropriate choice of other ligands.

Examples. (i) What is pPb for a system 0.1M in Mg^{2+}, 0.01M in EDTA and 0.001M in total lead?

From Table 7.3, log β^{MgEDTA} = 8.7 and log β^{PbEDTA} = 18.0. Substituting these values, $[M^{I}]_T = 0.001$, $[L]_T = 0.01$, and $[M^{II}]_T = 0.1$ in Equation 7.8 gives pPb = 11.3. (A more detailed analysis, using a computer and allowing for protonated metal-EDTA complexes, the pK_a values of EDTA and the hydrolysis constants of lead ion, gave values of pPb = 11.25 at pH 5.0, 11.29 at pH 7.0 and 11.29 at pH 9.0.)

(ii) In the above example, what would be the effect on pPb of replacing the magnesium by calcium or barium?

From Table 7.3, log β^{CaEDTA} = 10.6, giving pPb = 9.4. That is, the level of free lead ion would be raised if magnesium was replaced by calcium. Conversely, because log β^{BaEDTA} is only 7.7, replacement by barium would raise pPb to 12.3, so that the level of free lead ion would be lowered. This result can be generalized: the less the difference in stability constants of the two metal complexes in such buffer systems, the smaller will be pM (i.e. the higher will be the concentration) of the more strongly bonding metal ion.

7.5 Effects of pH buffer substances on pM

Although most of the chelating agents used in metal-ion buffers are also capable of serving as pH buffers, their pK_a values do not usually lie in the required pH range. For example the pK_a values of NTA, tartaric acid and citric acid are too remote from pH 7 for them to be used as pH buffers in near-physiological media. Similarly N-hydroxyethylethylenediamine triacetic acid (pK_a 5.33), EDTA (pK_a 6.13) or trimethylenediamine tetraacetic acid (pK_a 7.91) are of only limited use for this purpose.

Hence it is ordinarily necessary to control the pH of metal ion buffers by adding conventional buffering materials. Most of these, including acetate, borate, phosphate, bicarbonate and ammonia, have only weak metal-complexing ability by comparison with chelating agents. However, certain poly-dentate ligands such as the α-amino acids, α-hydroxy acids and dicarboxylic acids are quite strong chelating agents for metal ions and may significantly alter pM. As a useful guide, this effect can be ignored if pM as calculated for the metal with the acid-base buffer is at least two logarithm units less than pM for the metal buffer system. The metal-binding properties of some of the 'Good' buffers are summarized in Table 7.5.

The computer programmes COMICS (Perrin and Sayce, 1967) and HALTAFALL (Ingri *et al.*, 1967) are convenient if more detailed compositions of metal-complex mixtures have to be calculated. They also make possible the rapid screening of chelating agents that are potentially of use in this way.

For further reading on metal-ion buffers, see Chaberek and Martell (1959). Raaflaub (1956) described two examples of pCa-enzyme activity studies, and Wolf (1973) has very recently given a mathematical treatment of two-metal-ion buffer systems.

7.6 Anion buffers

An example of an anion buffer that may be suitable for standardizing an ion-selective electrode is based on the greater insolubility of silver iodide than silver chloride (Havas *et al.*, 1971). This iodide ion buffer comprises a solution containing a high concentration of free chloride ion to which is added a small amount of silver ion and a smaller amount of iodide ion. The chloride ion activity determines the activity of silver ion (because of the solubility product of AgCl) and, in turn, the activity of silver ion determines the activity of the iodide ion in the buffer solution. For chloride ion concentrations of 1M, 0.1M, and 0.01M, pI is approximately 5.85, 6.85 and 7.85, respectively.

Similarly, a pBr buffer should be formed by adding silver and bromide ions to a high concentration of chromate ion.

For a chromate concentration of 1M, 0.1M and 0.01M, pBr is approximately 6.4, 6.9 and 7.4, respectively. If iodate is used instead of chromate, pBr values become 4.8, 5.8 and 6.8. A $pCrO_4$ buffer might be obtained from a sulphate solution by adding lead and chromate ions. Sulphate concentrations of 1M, 0.1M and 0.01M give $pCrO_4$ equal to 4.8, 5.8 and 6.8, respectively.

Tanaka (1963) coined the term 'ligand buffers' for systems containing a ligand and an excess of metal ion. He pointed out that where a stable 1:1 chelate is formed, pL is determined only by the ratio of the total metal ion and total ligand concentrations and is independent of pH. The relation is

$$pL = \log \beta_1 + \log \left([M]_T - [L]_T\right)/[L]_T \qquad (7.9)$$

Although ligands are commonly anions, the treatment is a general one and applies equally to neutral ligands such as triethylenetetramine or tris(aminoethyl)amine.

7.7 Redox buffering

The concept of buffering can be extended to other physical processes. One of these is the buffering of a solution to maintain a constant oxidation-reduction potential. Placing a suitable reversible redox couple in a solution establishes and maintains the solution redox potential at an essentially constant value. Maximum buffer capacity should be found when the oxidant and reductant are present in equal concentrations. The technique was used by Kuntz *et al.*, (1964) to study photo-induced changes in bacterial chromophores. Examples included mixtures of $K_3 Fe(CN)_6$ and $K_4 Fe(CN)_6$ for the range $E = 0.30$ to 0.55 V, and oxidized and reduced formed of indigo tetra- and tri-sulphonic acids for $E = 0.02$ to -0.10 V (Reduction of the latter was controlled by addition of reduced indigodisulphonic acid prepared using sodium dithionite.)

Most of the reversible redox buffers are highly coloured

organic dyes: this may be a disadvantage, especially in biological studies. They may also have limited solubility or be chemically unstable. It is desirable to have at least a 10- to 100-fold excess of redox buffer material over the compound being investigated. Most biological redox potentials are pH-dependent, so that careful control of pH is essential.

For fuller details, see Swartz and Wilson (1971). Nightingale (1958) has discussed quantitative aspects.

Table 7.1 *Selected* pK_a values for hydrolysis of metal ions (mononuclear species only)*

Metal Ion	I	pK_1	pK_2	pK_3	pK_4
Ag^+	0	11.7	12.7		
Al^{3+}	0	5.0			
Ba^{2+}	0	13.4			
Be^{2+}	0.1	5.7	~ 7	10.5	
Bi^{3+}	3	1.6			
Ca^{2+}	0	12.6			
Cd^{2+}	3	9.9	10.6	11.6	12.3
Ce^{3+}		~ 9			
Ce^{4+}	2	-1.2	0.8		
Co^{2+}	0.1	8.9			
Co^{3+}	1	2.0			
Cr^{3+}	0	4.0	5.6		
Cu^{2+}	0	8.0			
Fe^{2+}	0.5	6.7			
Fe^{3+}	0.01	2.5	3.3		
Ga^{3+}	0	3.4			
Gd^{3+}	0.3	8.4			
Hf^{4+}	1	-0.1	0.2	0.4	0.5
Hg_2^{2+}	0.5	5.0			
Hg^{2+}	0.5	3.7	2.7		
In^{3+}	3	4.4	3.9		
La^{3+}	0.3	9.0			
Li^+	0	13.8			
Lu^{3+}	0.3	7.9			
Mg^{2+}	0	11.4			
Mn^{2+}	0	10.6			
Mn^{3+}	4	0.1			
Ni^{2+}	0	9.9			
Pb^{2+}	1	7.1	10.1	11.5	
Pd^{2+}	1	1.2			

Table 7.1 *(Cont.)*

Metal Ion	I	pK_1	pK_2	pK_3	pK_4
Sc^{3+}	1	5.1			
Sn^{2+}	1	1.7			
Sr^{2+}	0	13.2			
Th^{4+}	1	4.3	4.2		
Ti^{3+}	0	1.3			
TiO^{2+}		0.3			
Tl^+	0	13.2			
Tl^{3+}	3	1.1	1.5		
U^{4+}	0.2	1.1			
$UO_2{}^{2+}$	0.5	5.7			
V^{3+}	1	2.9	3.9		
VO^{2+}		5.4			
Y^{3+}	0.3	8.3			
Zn^{2+}	0	9.0			
Zr^{4+}	1	−0.3	0.1	0.4	0.6

*from Perrin (1969).

Table 7.2 *pK Values of some ligands useful as metal-ion buffers*
(at $25°C$ and $I = 0.15$)

Ligand	pK_a Values
Citric acid (H_3L)	5.62, 4.34, 2.91
EDTA (H_4L)	10.27, 6.16, 2.67, 2.00
NTA (H_3L)	9.75, 2.49, 1.89
Tartaric acid (H_2L)	3.94, 2.88
DTPA (H_5L)	10.42, 8.76, 4.42, 2.56, 1.79
HIMDA	8.73, 2.2
N,N-Bishydroxyethylglycine	8.08

Abbreviations: EDTA = ethylenediaminetetraacetic acid;
NTA = nitrilotriacetic acid;
DTPA = diethylenetriaminepentaacetic acid;
HIMDA = N-Hydroxyethyliminodiacetic acid.

Table 7.3 *Stability constants* of metal-complexes*

(at $25°C$ and $I = 0.15$)

Ligand	Metal ion	Log β_1	Other log βs
Citric acid	Ba^{2+}	2.9	$K(Ba^{2+} + HL^{2-} = BaHL)$ 1.75
	Ca^{2+}	3.6	$K(Ca^{2+} + HL^{2-} = CaHL)$ 2.1
	Cd^{2+}	3.8	$K(Cd^{2+} + HL^{2-} = CdHL)$ 2.2
	Co^{2+}	5.0	$K(Co^{2+} + HL^{2-} = CoHL)$ 3.0
	Cu^{2+}	5.9	$K(Cu^{2+} + HL^{2-} = CuHL)$ 3.4
	Fe^{3+}	11.4	$K(2Fe^{3+} + 2L^{3-}$ $= Fe_2(H_{-1}L)_2^{2-} + 2H^+)$ 21.2
	Mg^{2+}	3.4	$K(Mg^{2+} + HL^{2-} = MgHL)$ 1.8
	Mn^{2+}	3.7	
	Ni^{2+}	5.4	$K(Ni^{2-} + HL^{2-} = NiHL)$ 3.3
	Zn^{2+}	5.0	$K(Zn^{2+} + HL^{2-} = ZnHL)$ 3.0
EDTA	Ag^+	7.3	$K(Ag^+ + HL = AgHL)$ 3.4
	Ba^{2+}	7.7	
	Ca^{2+}	10.6	
	Cd^{2+}	16.6	
	Co^{2+}	16.5	
	Cu^{2+}	18.9	
	Fe^{3+}	25.1	
	Hg^{2+}	21.8	
	Mg^{2+}	8.7	
	Mn^{2+}	14.1	
	Mn^{3+}	24.8	
	Ni^{2+}	18.7	
	Pb^{2+}	18.0	
	Zn^{2+}	16.7	
NTA	Ag^+	5.2	
	Ba^{2+}	4.8	
	Ca^{2+}	6.5	
	Cd^{2+}	10.0	β_2 (15.5)
	Co^{2+}	10.8	β_2 (14.3)
	Cu^{2+}	13.1	
	Fe^{3+}	15.9	β_2 (24.6)
	Hg^{2+}	14.6	
	Mg^{2+}	5.5	
	Mn^{2+}	7.4	
	Ni^{2+}	11.5	
	Pb^{2+}	11.5	
	Zn^{2+}	10.4	
Tartaric acid	Ba^{2+}	2.2	
	Ca^{2+}	2.2	
	Co^{2+}	2.3	
	Fe^{3+}	6.5	β_2 (11.9)
	Mg^{2+}	1.9	

Table 7.3 *(Cont.)*

Ligand	Metal ion	Log β_1	Other log βs
	Mn^{2+}	1.4	
	Ni^{2+}		β_2 (5.4)
	Pb^{2+}	2.9	
	Zn^{2+}	3.1	β_2 (5.0)
DTPA	Ag^+	8.7	
	Ba^{2+}	8.6	
	Ca^{2+}	10.6	
	Cd^{2+}	19.1	
	Co^{2+}	19.0	
	Cu^{2+}	21.1	β_2 (27.9)
	Fe^{3+}	28.6	
	Hg^{2+}	28.4	
	Mg^{2+}	9.3	
	Mn^{2+}	15.1	
	Ni^{2+}	20.2	
	Pb^{2+}	18.8	
	Zn^{2+}	18.7	
HIMDA	Ba^{2+}	3.4	
	Ca^{2+}	4.7	
	Cd^{2+}	7.3	β_2 (12.5)
	Co^{2+}	8.1	β_2 (12.4)
	Cu^{2+}	11.9	β_2 (16.0)
	Fe^{2+}	6.8	β_2 (10.0)
	Fe^{3+}	11.6	
	Hg^{2+}	5.5	β_2 (8.8)
	Mg^{2+}	3.5	
	Mn^{2+}	5.6	β_2 (9.4)
	Ni^{2+}	9.4	β_2 (14.5)
	Pb^{2+}	9.5	β_2 (13.6)
	Sr^{2+}	3.8	
	Zn^{2+}	8.4	β_2 (12.3)
N,N-Bishydroxy-ethylglycine	Cd^{2+}	4.8	β_2 (8.2)
	Co^{2+}	5.3	β_2 (8.8)
	Cu^{2+}	8.1	β_2 (13.3)
	Fe^{2+}	4.3	β_2 (7.3)
	Mn^{2+}	3.1	β_2 (5.4)
	Ni^{2+}	6.4	β_2 (10.8)
	Zn^{2+}	5.4	β_2 (8.6)

*Averaged values taken from Sillén & Martell (1964, 1971).

Table 7.4 *Some representative pM levels using two-metal-ion EDTA buffer systems for pH range 6–10*

$[Ca]_T$ (M)		$[Mg]_T$ (M)	[EDTA] (M)	pCa
0.001		0.034	0.004	3.9
0.01		0.03	0.02	3.9
0.0001		0.01	0.001	4.9
0.001		0.021	0.011	4.9
0.0001		0.02	0.01	5.9
0.0001		0.03	0.02	6.2
0.0001		0.06	0.05	6.6
0.0001		0.10	0.09	6.9

$[Zn]_T$ (M)	M^{II}	$[M^{II}]_T$ (M)	[EDTA] (M)	pZn
0.001	Mn(II)	0.10	0.01	4.6
0.001	Mn(II)	0.021	0.011	5.6
0.001	Mn(II)	0.041	0.031	6.1
0.001	Mn(II)	0.046	0.041	6.5
0.01	Ca	0.05	0.02	7.6
0.01	Ca	0.10	0.04	7.8
0.001	Ca	0.10	0.01	8.1
0.001	Ca	0.041	0.011	8.6
0.005	Mg	0.05	0.01	9.4
0.001	Mg	0.10	0.01	10.0
0.02	Mg	0.10	0.09	10.5
0.001	Mg	0.021	0.013	11.0

Table 7.5 *Metal-binding abilities of some of the newer buffers**

Buffer	pK	Ca	log β_1 Cu	Mg	Mn
ACES	6.9	0.4	4.6	0.4	—
ADA	6.6	4.0†	9.7†	2.5†	4.9†
Bicine	8.35	2.8	8.1	1.5	3.1
Glycylglycine	8.4	0.8	5.8	0.8	1.7
MES	6.15	0.7	—	0.8	0.7
Tricine	8.15	2.4	7.3	1.2	2.7

*Good *et al.* (1966)
†Schwarzenbach *et al.* (1955)

Chapter Eight

Purification of Substances used in Buffers

Many of the components of buffer mixtures are commercially available in analytical reagent grade and, after suitable drying, may be used directly in distilled or de-ionized water. The drying procedure is important because many substances are hygroscopic. When substances are known to deteriorate on storage, when there is evidence of decomposition, or when known impurity levels in reagents are unacceptable, further purification is required. Where possible, the strengths of solutions prepared from substances of doubtful purity or degree of hydration should be confirmed by titration.

Details are available for removing undesirable metal ions in the preparation of acids, alkalis and buffer solutions of high purity. The methods are based on exhaustive extraction with dithizone and on isopiestic distillation of ammonia and hydrochloric acid (Irving and Cox, 1958).

A problem in the use of many amines as pH buffers is that they are liquids or gases at room temperature and are difficult to obtain pure. Of the solid amines, 4-amino-pyridine, imidazole and Tris are ordinarily available in acceptable purity, but others, such as 2-amino-2-methyl-1,3-propanediol, 2-amino-2-methyl-1-propanol and hexamethylenediamine, may need further purification, commonly by distillation. However it is difficult to remove residual monoethanolamine and diethanolamine from triethanolamine. A general method for the purification of amines is through the preparation of their hydrochlorides.

The amine is dissolved in a suitable solvent (typically benzene) and dry HCl gas is passed into the solution to precipitate the amine hydrochloride which is purified by recrystallization from a suitable solvent mixture (typically benzene/ethanol). Buffer solutions are prepared by taking a weighed quantity of the hydrochloride and adding standard sodium hydroxide solution. If desired, the free amine can be regenerated from the hydrochloride by adding solid sodium hydroxide.

Procedures, taken from the literature, are outlined below for purifying some commonly used buffer substances. Where purification details are lacking for particular compounds, reference to Perrin *et al.*, (1966) may be helpful.

N-(2-Acetamido)-2-aminoethanolsulphonic acid, (ACES), (M.Wt.182.20). Recryst. from hot aq. ethanol.

N-(2-Acetamido)iminodiacetic acid, (ADA), (M.Wt. 190.16). Dissolve in water by adding one equivalent of sodium hydroxide solution (to final pH of 8–9), then acidify with hydrochloric acid to precipitate the free acid. Filter and wash.

Acetic acid, (M.Wt. 60.05). Very hygroscopic. Water and other impurities can be removed by adding excess acetic anhydride and chromic trioxide (2 g 100 ml^{-1}), heating to near boiling point (118°C) for an hour, and then fractionally distilling. An alternative purification uses fractional freezing of glacial acetic acid (pure material freezes at 16.6°C).

trans-Aconitic acid, (M.Wt. 174.11). Recryst. from water. (Solubility at 25°C is 0.5 g ml^{-1}). Dry in a vacuum desiccator.

(2-Aminoethyl)trimethylammonium chloride hydrochloride, (Cholamine chloride hydrochloride), (M.Wt. 175.11). Recryst. from ethanol. (Material very soluble in water.)

2-Amino-2-methyl-1,3-propanediol, (M.Wt. 105.14, m.p. 111°C). Recryst. from methanol. Dry in a stream of dry nitrogen at room temperature, then in a vacuum oven at 55°C. Store over calcium chloride.

4-Aminopyridine, (M.Wt. 94.12, m.p. 160°C). Cryst. from benzene-ethanol, then twice from water. Dry at 105°C.

Ammonium chloride, (M.Wt. 53.50). Cryst. from water (0.7 g ml^{-1}).

N,N-Bis(2-hydroxyethyl)-2-aminoethanesulphonic acid, (BES), (M.Wt. 213.25). Recryst. from aq. ethanol.

N,N'-Bis(2-hydroxyethyl)glycine, (Bicine), (M.Wt. 163.18). Recryst. from 80% methanol.

[Bis-(2-hydroxyethyl)imino]-tris-[(hydroxymethyl)methane], *(Bis-tris)*, (M.Wt. 209.25, m.p. 89°C). Recryst. from hot 1-butanol. Dry in vac. at 25°C.

Boric acid, (M.Wt. 61.83). Recryst. three times from boiling water (0.3 g ml^{-1}). Filter the hot solution through sintered glass. Cool to 0°C. Air-dry at room temperature.

Cacodylic acid, $(CH_3)_2AsO_2H$, (M.Wt. 137.99). Cryst. from warm ethanol. Dry in vac. desiccator over $CaCl_2$.

Calcium hydroxide, (M.Wt. 74.09). Heat analytical grade calcium carbonate to 1000°C during one hour. Allow the resulting oxide to cool and add slowly to water. Heat the suspension to boiling, cool and filter through a sintered glass funnel of medium porosity (to remove soluble alkaline impurities). Dry the solid at 110°C and crush to a uniformly finely granular condition.

Chloroacetic acid, (M.Wt. 94.50). Recryst. from benzene or CCl_4. Dry in a vac. desiccator over conc. sulphuric acid or P_2O_5.

Citric acid (monohydrate), (M.Wt. 210.14). Cryst. from hot water.

s-Collidine, see 2,4,6-Trimethylpyridine.

Diethanolamine, (M.Wt. 105.14, m.p. 28°C, b.p. 154–155°C 10 torr^{-1}). Vacuum distil twice. Final purification by fractional crystn. from its melt.

5,5-Diethylbarbituric acid, (M.Wt. 184.20, m.p. 188–192°C). Cryst. from ethanol. Dry in a vac. desiccator over P_2O_5.

Diethylenetriaminepentaacetic acid, *(DTPA)*, (M.Wt. 393.35). Recryst. from water. Dry under vac. or at 110°C.

3,3-Dimethylglutaric acid, (M.Wt. 160.17, m.p. 103–104°C). Cryst. from water. Dry under vacuum.

Disodium hydrogen orthophosphate (anhydrous), (M.Wt. 141.96). Cryst. twice from warm water. Dry at 110–130°C for 2 hours. (The salt is hygroscopic and should be dried before use.)

Ethanolamine, (M.Wt. 61.08, m.p. 10.5°C, b.p. 171.1°C). Distil at low pressure (5 torr). Can be cryst. by cooling, alone or as a solution in ethanol.

Ethanolamine hydrochloride, (M.Wt. 97.54). Cryst. from ethanol. (Salt is deliquescent.)

Ethylenediamine, (M.Wt. 60.10, b.p. 117.0°C). Dry with solid NaOH or molecular sieve. Distil under nitrogen. Protect from carbon dioxide.

Ethylenediaminetetraacetic acid, (M.Wt. 292.25). Dissolve in aq. ammonia or KOH and precipitate with dilute HCl or HNO_3. Repeat. Boil with distilled water, twice, to remove mineral acid, then recryst. from water or N,N-dimethylformamide. Dry at 110°C.

(The disodium salt of EDTA is commercially available as a dihydrate (M.Wt. 372.24) which, if dried at 80°C to give the anhydrous form, is suitable as an analytical standard. If less pure material is available the salt can be recryst. from water or precipitated from a 10% aqueous solution by the addition of ethanol, and then dried at 80°C.)

N-Ethylmorpholine, (M.Wt. 115.18, b.p. 138°C). Distil before use.

Furoic acid, (M.Wt. 112.09, m.p. 131–132°C). Cryst. from hot water, dry at 120°C, then cryst. from $CHCl_3$ and again dry.

Glycinamide hydrochloride, (M.Wt. 110.54). Cryst. from ethanol.

Glycine, (M.Wt. 75.07). Dissolve in warm water (0.5 g ml^{-1}) and cryst. by addition of ethanol or methanol. Wash with the alcohol, then ethyl ether. Dry at 110°C.

N-Glycylglycine, (M.Wt. 132.12). Cryst. from aq. ethanol (1:1). Dry at 110°C.

2-Hydroxyethylimino-tris(hydroxymethyl)methane,
(Mono-tris), (M.Wt. 165.19, m.p. 91°C). Cryst. twice from ethanol. Dry under vacuum at 25°C.

N-2-Hydroxyethylpiperazine-N'-2-ethanesulphonic acid, (HEPES), (M.Wt. 238.31). Cryst. from hot ethanol and water.

Hydroxylamine hydrochloride, (M.Wt. 69.49). Cryst. from boiling methanol. Dry under vacuum over P_2O_5.

Imidazole, (M.Wt. 68.08, m.p. 89.5−90°C). Cryst. from benzene, CCl_4 or petroleum ether. Dry at 40°C under vacuum.

Maleic acid, (M.Wt. 116.07, m.p. 143.5°C). Cryst. from hot water. Dry at 110°C.

2-(N-Morpholino)ethanesulphonic acid, (MES), (M.Wt. 195.24). Cryst. from hot ethanol containing a little water.

Nitrilotriacetic acid, (NTA), (M.Wt. 191.14). Cryst. from water. Dry at 110°C.

Phenylacetic acid, (M.Wt. 136.15, m.p. 76−77°C). Cryst. from aq. ethanol. Dry under vac.

Piperazine, (M.Wt. 86.14, m.p. 110−112°C). Cryst. from ethanol. Dry under vacuum.

Piperazine-N,N'-bis(2-ethanesulphonic acid), (PIPES), (M.Wt. 302.37). Cryst. from boiling water (maximum solubility is about $1 \, g \, l^{-1}$) or as described for ADA.

Piperazine-N,N'-bis(2-ethanesulphonic acid), monosodium salt, (monohydrate), (M.Wt. 342.4). Cryst. from water and ethanol.

Piperazine dihydrochloride, (M.Wt. 159.06). Cryst. from aq. ethanol. Dry at 110°C.

Piperazine phosphate (monohydrate), $C_4H_{10}N_2 \cdot H_3PO_4 \cdot H_2O$, (M.Wt. 202.15). Prepared by adding an equimolar mixture of piperazine and phosphoric acid in aqueous solution. Cryst. twice from water, air-dried and stored for several days over Drierite. Salt dehydrates slowly if heated at 70°C.

Potassium dihydrogen citrate, (M.Wt. 230.22). Cryst. from water. Dry at 80°C.

Potassium dihydrogen orthophosphate (anhydrous), (M.Wt. 136.09). Cryst. from water, then ethanol, and again from water. Dry at 110°C for 1−2 hours.

Potassium hydrogen phthalate, (M.Wt. 204.22). Cryst. from water. Dry at 110°C.

Potassium hydrogen tartrate, (M.Wt. 188.18). Cryst. from boiling water (.06 $g \, ml^{-1}$). Dry at 110°C.

Potassium p-phenolsulphonate, $C_6H_4(OH)SO_3K$, (M.Wt. 212.27). Cryst. several times from hot water, adding charcoal if necessary. Dry at 110°C for 24 hours.

Potassium tetroxalate, $KH_3(C_2O_4)_2 \cdot 2H_2O$, (M.Wt. 254.20). Cryst. from water, taking precautions that crystals do not separate above 50°C. Dry below 50°C. (Slight variations in the water content of different batches will not significantly affect its use as a primary pH standard.)

Pyridine, (M.Wt. 79.10, b.p. 115.6°C). Dry by refluxing with solid KOH, then fractionally distil.

Sodium acetate (trihydrate), (M.Wt. 136.08). Cryst. from aq. ethanol.(The anhydrous salt can be prepared by heating the trihydrate slowly till it liquefies, steam is evolved and the mass again solidifies. Heating is increased slowly, to avoid charring, until the salt again fuses. After several minutes it is allowed to cool, is powdered and bottled.)

Sodium bicarbonate, (M.Wt. 84.01). Cryst. from hot water (6 ml/g). Dry below 40°C.

Sodium borate (decahydrate), (borax), (M.Wt. 381.38). Cryst. from water (3.3 ml/g), keeping the temperature below 55°C (to avoid formation of the pentahydrate). Wash with water. Equilibrate for several days in a desiccator containing an aqueous solution saturated with respect to both sucrose and sodium chloride. (Less reliably, because of variable water content in the final product, washing with water can be followed by ethanol and ether, with air-drying at room temperature for 12−18 hours on a clock glass.)

Sodium cacodylate (trihydrate), $(CH_3)_2AsO_2Na \cdot 3H_2O$, (M.Wt. 214.03). Cryst. from aq. ethanol.

Sodium carbonate (decahydrate), (M.Wt. 286.15). Cryst. from water. Anhydrous sodium carbonate is prepared by heating the decahydrate or sodium bicarbonate at 270°C for 2 hours.

Sodium 5,5-diethylbarbiturate, (M.Wt. 206.18). Cryst. from warm water (3 ml/g) by adding an equal volume of ethanol and cooling. Dry under vacuum over P_2O_5.

Sodium dihydrogen orthophosphate (dihydrate), (M.Wt. 156.01). Cryst. from warm water (0.5 ml/g).

Sodium hydrogen diglycollate, $NaOOCCH_2OCH_2COOH$, (M.Wt. 156.08). Prepared from diglycollic acid (600 g in 1 litre of distilled water) by adding 1 equivalent of NaOH as 4–5M solution. (The required amount is estimated from half the titration to neutrality of an aliquot of the free acid, against phenolphthalein.) After cooling to room temperature the precipitate is filtered off and recryst. twice from hot water, then dried overnight at 110°. (At room temperature the solubility is about 5 g 100 ml^{-1}.)

Sodium pyrophosphate (decahydrate), (M.Wt. 446.1). Cryst. from water. Air-dried at room temperature.

Sodium succinate (anhydrous), (M.Wt. 162.06). Cryst. from water (1.2 ml/g). Dry at 125°C.

Sodium p-toluenesulphonate, (M.Wt. 194.19). Dissolve in water, filter, evaporate to dryness. Recryst. from ethanol. Dry at 110°C.

Succinic acid, (M.Wt. 118.09). Cryst. from water. Dry over concentrated H_2SO_4 in a desiccator.

D(+)-Tartaric acid, (M.Wt. 150.09, m.p. 169.5–170°C). Cryst. from water.

p-Toluenesulphonic acid (monohydrate), (M.Wt. 190.22). Pass HCl gas through a saturated aqueous solution at 0°C. Dry in a vac. desiccator over solid KOH or $CaCl_2$. (Solid is somewhat deliquescent.)

Trichloroacetic acid, (M.Wt. 163.39). Recryst. several times from dry benzene. Store over concentrated H_2SO_4 in a vac. desiccator.

Triethanolamine hydrochloride, (M.Wt. 185.65). Cryst. from ethanol. Dry at 80°C.

2,4,6-Trimethylpyridine, (2,4,6-Collidine), (M.Wt. 121.18). (The commercial material may be grossly impure.) Dissolve in methanol and add 85% phosphoric acid, with cooling, to precipitate the amine phosphate. (Alternatively, dry HCl gas can be passed into a benzene/2,4,6-trimethylpyridine mixture, cooled in an ice-bath.) Filter off the salt, regenerate the free base by adding aq. NaOH and extract with benzene. Dry with anhydrous magnesium sulphate and distil under reduced pressure (b.p. 70°C 13 torr^{-1}).

Tris(hydroxymethyl)aminomethane, *(Tris),* *(THAM)* (M.Wt. 121.14, m.p. 169.5–171.5°C). Tris can ordinarily be obtained in highly pure form suitable for use as an acidimetric standard. If only impure material is available, cryst. from 20% aq. ethanol. Dry in a vac. desiccator over P_2O_5 or $CaCl_2$.

Tris(hydroxymethyl)aminomethane hydrochloride, (Tris-HCl), (M.Wt. 157.60). Tris-hydrochloride is also available commercially in a highly pure state. Otherwise, cryst. from 50% ethanol, then from 70% ethanol and dry below 40°C to avoid risk of decomposition.

N-Tris(hydroxymethyl)methyl-2-aminoethanesulphonic acid, (TES), (M.Wt. 229.25). Cryst. from hot ethanol containing a little water.

N-Tris(hydroxymethyl)methylglycine, *(Tricine),* (M.Wt. 179.17). Cryst. from ethanol and water.

Trisodium citrate (dihydrate), $Na_3C_6H_5O_7 \cdot 2H_2O$ (M.Wt. 294.10). Cryst. from warm water.

Trisodium orthophosphate (dodecahydrate), (M.Wt. 380.14). Cryst. from warm dilute aq. NaOH by cooling to 0°C.

Chapter Nine

Preparation of Buffer Solutions

Materials used in the preparation of buffer solutions should be good quality laboratory chemicals, purified if necessary as described in Chapter 8 and dried to constant composition. The distilled water used as solvent should have been recently boiled to remove dissolved carbon dioxide and have been protected from contamination by atmospheric carbon dioxide while cooling. (This precaution is unnecessary in preparing buffer solutions having pH values less than 5.) The water should have a specific conductivity of less than 2×10^{-6} ohm^{-1} cm^{-1} at 25°C. Solutions should be stored in stoppered Pyrex (or similar borosilicate glass) or pure polyethylene bottles.

It is important that specified concentrations be adhered to in preparing buffer solutions. Attempts to use stronger or more dilute solutions than specified will introduce error, even when the concentration ratio of the buffer species remains constant. This is because of changes in the ionic strength and, consequently, a change in the practical pK_a' value of the buffer acid. Where buffer solutions must be diluted, the effect on the pH of the solution can be calculated to within a few hundredths of a pH unit from the Debye-Hückel equation, as described in Chapter 2.

Amine buffers are usually prepared from a weighed or measured quantity of an amine or its solution by adding a measured quantity of a standard solution of a strong acid which neutralises between 25% and 75% of the amine. For amine buffers of constant ionic strength it is convenient to mix solutions of the amine hydrochloride containing added sodium chloride with standard alkali containing added sodium chloride. For an ionic strength, I, in the final buffer mixture the solutions might be:

(I) 0.05M amine-HCl + $(I - 0.05)$M NaCl
(II) 0.20M NaOH + IM NaCl.

By appropriately varying the volume ratios of these solutions the pH of the buffer system can be varied over the useful buffer range while maintaining a constant ionic strength. (The total moles of NaOH must not exceed the amount of amine hydrochloride present.)

The preparation of solutions for metal-ion buffers is straightforward. The ligand should be chemically pure, stable, water-soluble and not deliquescent. If the buffers are for use in biochemical or biological experiments they should not be toxic. It is convenient to prepare two stock solutions. One of these contains a known concentration of the metal chelate (usually by mixing equal amounts of the sodium or potassium salt of the chelating agent with a soluble salt of the metal ion, and adding alkali dropwise to adjust the pH to the desired value). The other is the metal-free ligand adjusted to the desired pH with alkali. Mixing the two solutions in varying proportions provides a ready means to vary pM. For calcium or magnesium ion buffers, pH constancy can be achieved by adding acetate, succinate, imidazole, barbiturate or most of the 'Good' buffers. With heavy metal ions the choice is more restricted and it may be necessary to allow for the effects of binding by the pH buffer material, using equations given in Chapter 7.

For alkaline solutions, including amines, borax and some phosphates, it is necessary to guard against absorption of carbon dioxide. Detailed procedures for preparing and handling carbon dioxide-free solutions of alkalis are described by Albert and Serjeant (1971).

The primary buffer standards – phthalate, phosphate and borax solutions – are prepared by direct solution of weighed amounts in water and dilution to a known volume. Solutions of acid or alkali used in the preparation of buffers should be standardized by titration. Even so, *to guard against any inadvertent error, the user is strongly advised to check with a pH meter the pH values of buffer solutions at the temperature for which they are required.*

Except for tartrate solutions, buffers can be stored

indefinitely in a cool place but, as a general procedure, it is preferable to replace them regularly once a month.

The following notes on individual components of buffer solutions may be helpful.

Acetic acid, CH_3COOH, (M. Wt. 60.05). To prepare approx. 1M solution, dilute 57.5 ml of pure glacial acetic acid to 1 litre. Check by titration with alkali against phenolphthalein. Acetic acid/sodium acetate buffers are prepared by partial neutralization of acetic acid solutions with NaOH : NaOAc should not be used.

2-Amino-2-methylpropane-1,2-diol, $C_4H_{11}O_2N$, (M. Wt. 105.14). To prepare a 1M solution, dissolve 105.14 g of the free base in 1 litre of CO_2-free water. Check by titration with HCl against methyl orange.

Ammonia, NH_3, (M. Wt. 17.03). To prepare approx. 1M solution, dilute 60 ml of s.g. 0.88 or 72 ml of s.g. 0.91 concentrated ammonia solution to 1 litre with CO_2-free water. Check by titration with HCl against methyl orange.

Borax. See Sodium tetraborate.

Cacodylic acid, $(CH_3)_2AsO_2H$, (M. Wt. 137.99). To prepare a 1M solution dissolve 137.99 g in 1 litre. Check by titration with NaOH against phenolphthalein.

Calcium hydroxide, $Ca(OH)_2$, (M. Wt. 74.10). The preparation of $Ca(OH)_2$ is described in Chapter 8. A saturated solution of $Ca(OH)_2$ at $25°C$ (0.02025M) is obtained by shaking upwards of 1.5 g $Ca(OH)_2$ with 1 litre of water and allowing to stand so that undissolved material can settle out and be removed by filtration (suction) through a medium porosity sintered funnel. Its pH at $25°C = 12.454$; $d(pH)/dT = -0.033$.

5.5-Diethylbarbituric acid, *(Barbitone)*, $C_8H_{12}N_2O_3$, (M. Wt. 184.20). A saturated solution is 0.017M at $0°C$, 0.019M at $5°C$ and 0.028M at $25°C$. To prepare an 0.025M solution, dissolve 4.605 g in 1 litre.

Disodium hydrogen phosphate, Na_2HPO_4, (M. Wt. 141.97). This substance is hygroscopic and should be dried at $110°C$ for $1-2$ hours before use. Mixtures with KH_2PO_4 are used as primary pH standards. Buffers may slowly develop sediment by leaching calcium from glass. To prepare an 0.5M solution, dissolve 70.99 g in CO_2-free water and dilute to 1 litre.

Formic acid, HCOOH, (M. Wt. 46.03). To prepare an approx. 1M solution, dilute 38.5 ml '98−100%' formic acid or 42.4 ml '90%' formic acid to 1 litre. Check by titration with NaOH against phenolphthalein. Formic acid/sodium formate buffers are prepared by partial neutralization of formic acid solutions with NaOH.

Glycine, NH_2CH_2COOH, (M. Wt. 75.07). Dry at 100°C for 1−2 hours before use.

Hydrochloric acid, HCl, (M. Wt. 36.47). To prepare approx. 1M solution dilute 90 ml conc. HCl ('36% HCl') to 1 litre. Check by titration with standard alkali against methyl red. Weaker concentrations can be prepared by dilution.

Potassium chloride, KCl, (M. Wt. 74.56). Dry at 110°C or higher for several hours before weighing.

Potassium dihydrogen phosphate, KH_2PO_4 (M. Wt. 136.09). Dry at 110°C for 1−2 hours before use. See notes under *Disodium hydrogen phosphate.*

Potassium hydrogen phthalate, $KHC_8H_4O_4$, (M. Wt. 204.22). Dry at 110°C for 1−2 hours before use. Cool in a desiccator over $CaCl_2$. This is a primary pH standard, 0.05M = 10.212 g litre^{-1} has pH 4.008 at 25°C, $d(pH)/dT$ = 0.0012. If high accuracy is required, concentrations can be converted to the molal scale by taking the density as 1.0017. The buffer does not absorb CO_2 but may slowly develop mould.

Potassium hydrogen tartrate, $KHC_4H_4O_6$, (M. Wt. 188.18). Dry at 110°C for 1−2 hours before use. A saturated solution at 25°C (0.034M = 6.398 g litre^{-1}) is a primary pH standard, pH = 3.557, $d(pH)/dT$ = −0.0014. The buffer solution is subject to mould growth, leading to an increase in pH, and should be made freshly every few days or be preserved with a few crystals of thymol.

Potassium 4-phenolsulphonate, $C_6H_4(OH)SO_3K$, (M. Wt.

212.27). Dry at 110°C before use. To prepare a 1M solution, dissolve 212.27 g in CO_2-free water and dilute to 1 litre. Solutions must be kept above 20°C or solid will separate.

Potassium tetroxalate, $KH_3(C_2O_4)_2 \cdot 2H_2O$, (M. Wt. 254.20). Dry below 60°C before use. An 0.05M solution (0.0496M = 12.61 g litre^{-1}) is a secondary pH standard, pH at 25°C = 1.679, $d(pH)/dT = 0.001$.

Sodium cacodylate, $(CH_3)_2AsO_2Na \cdot 3H_2O$, (M. Wt. 214.02). To prepare a 1M solution, dissolve 214.02 g in CO_2-free water and dilute to 1 litre. Check by titration with HCl against methyl orange.

Sodium carbonate, Na_2CO_3, (M. Wt. 105.99). The salt is slightly hygroscopic and should be dried by heating strongly (it is stable up to 280°C) before use.

Sodium 5,5-diethylbarbiturate, (*Sodium barbitone*), $C_8H_{11}N_2O_3Na$, (M. Wt. 206.18). To prepare an 0.5M solution, dissolve 103.09 g in CO_2-free water and dilute to 1 litre. Check by taking an aliquot, diluting 20-fold, and titrating with HCl against methyl orange. The solubility is 0.79M at 5°C, 0.82M at 25°C. Solutions decompose slowly at room temperature, more rapidly if they are kept warm.

Sodium hydrogen maleate, $C_4H_3O_4Na$, (M. Wt. 128.06). To prepare an 0.2M solution, dissolve 23.22 g maleic acid or 19.61 g maleic anhydride in aqueous NaOH so that the final volume of 1 litre is 0.2M in sodium ion.

Sodium hydroxide, NaOH (M. Wt. 40.00). For the preparation of standard solutions of CO_2-free NaOH, and for precautions in storage, see Albert and Serjeant, (1971).

Sodium tetraborate, $Na_2B_4O_7 \cdot 10H_2O$, (M. Wt. 381.42). The solid material should not be heated above room temperature as it readily loses water. An 0.01M solution (3.814 g litre^{-1}) is a primary pH standard, pH at 25°C = 9.180, $d(pH)/dT = -0.0082$. To avoid contamination with CO_2, the storage bottle should be stoppered except when in use, or protected with a soda-lime tube. Use buffer solutions within ten minutes of removal from the bottle.

2,4,6-Trimethylpyridine, $C_8H_{11}N$, (M. Wt. 121.18). To prepare an approx. 0.2M solution, dissolve 27 ml in CO_2-free water and dilute to 1 litre. Check by titration with HCl against methyl orange.

Tris(hydroxymethyl)aminomethane, $C_4H_{11}O_3N$, (M. Wt. 121.14). Dry under vacuum at 70°C. To prepare a 1M solution, dissolve 121.14 g in CO_2-free water and dilute to 1 litre. Check by titration with HCl against methyl orange.

Tris(hydroxymethyl)aminomethane acid maleate, $C_8H_{15}O_7N$, (M. Wt. 237.22). To prepare an 0.2M solution, dissolve 24.23 g Tris and 23.22 g maleic acid or 19.61 g maleic anhydride in water and dilute to 1 litre.

Trisodium citrate dihydrate, $Na_3C_6H_5O_7 \cdot 2H_2O$, (M. Wt. 294.11). To prepare an 0.1M solution, dissolve 29.41 g in CO_2-free water and dilute to 1 litre. The pentahydrate should not be used.

Chapter Ten

Appendices

Appendix I:
Tables for constructing buffer tables

Tables 10.1 and 10.2 enable buffer tables to be constructed for use at a constant ionic strength of 0.1 for monoacidic bases or monobasic acids, given the appropriate 'practical' pK_a'. Where necessary, this can be calculated from the thermodynamic value in Appendix III by using Table 2.3.

In Table 10.3, a pH-composition table has been computed for buffers 0.05M in monoacidic base, adjusted by adding 0.1M HCl. Similarly, Table 10.4 is for buffers 0.05M in monobasic acid, adjusted by adding 0.1M NaOH.

Table 10.1 *Compositions of monoacidic base buffers at constant ionic strength (1 = 0.1)*

pH	Titre*	β†	NaCl‡
pH = pK_a' − 1.00	45.5	0.0095	0.45
− 0.95	45.0	0.0104	0.50
− 0.90	44.4	0.0114	0.56
− 0.95	43.8	0.0125	0.62
− 0.80	43.2	0.0136	0.68
− 0.75	42.5	0.0148	0.75
− 0.70	41.7	0.0160	0.83
− 0.65	40.9	0.0172	0.91
− 0.60	40.0	0.0185	1.00
− 0.55	39.0	0.0197	1.10
− 0.50	38.0	0.0210	1.20
− 0.45	36.9	0.0223	1.31
− 0.40	35.8	0.0235	1.42
− 0.35	34.6	0.0246	1.54
− 0.30	33.3	0.0256	1.67
− 0.25	32.0	0.0265	1.80
− 0.20	30.7	0.0273	1.93
− 0.15	29.3	0.0279	2.07
− 0.10	27.9	0.0284	2.21

Table 10.1 (*Cont.*)

pH	Titre*	$\beta\dagger$	NaCl‡
− 0.05	26.4	0.0287	2.36
− 0.00	25.0	0.0288	2.50
+ 0.05	23.6	0.0287	2.64
+ 0.10	22.1	0.0284	2.79
+ 0.15	20.7	0.0279	2.93
+ 0.20	19.3	0.0273	3.07
+ 0.25	18.0	0.0265	3.20
+ 0.30	16.7	0.0256	3.33
+ 0.35	15.4	0.0246	3.46
+ 0.40	14.2	0.0235	3.58
+ 0.45	13.1	0.0223	3.69
+ 0.50	12.0	0.0210	3.80
+ 0.55	11.0	0.0197	3.90
+ 0.60	10.0	0.0185	4.00
+ 0.65	9.1	0.0172	4.09
+ 0.70	8.3	0.0160	4.17
+ 0.75	7.5	0.0148	4.25
+ 0.80	6.8	0.0136	4.32
+ 0.85	6.2	0.0125	4.38
+ 0.90	5.6	0.0114	4.44
+ 0.95	5.0	0.0104	4.50
+ 1.00	4.5	0.0095	4.55

*Ml of 0.1M HCl added to 10 ml 0.5M base. After addition of NaCl, diluted to give 100 ml (final solution) of 0.05M buffer.
†Buffer capacity.
‡Ml of 1M NaCl added to give final ionic strength of 0.1.

Table 10.2 *Composition of monobasic acid buffers at constant ionic strength* $(I = 0.1)$

	Titre*	$\beta\dagger$	NaCl‡
pH = $pK_a' − 1.00$	4.5	0.0095	9.55
− 0.95	5.0	0.0104	9.50
− 0.90	5.6	0.0114	9.44
− 0.85	6.2	0.0125	9.38
− 0.80	6.8	0.0136	9.32

Table 10.2   (*Cont.*)

	Titre*	β†	NaCl‡
− 0.75	7.5	0.0148	9.25
− 0.70	8.3	0.0160	9.17
− 0.65	9.1	0.0172	9.09
− 0.60	10.0	0.0185	9.00
− 0.55	11.0	0.0197	8.90
− 0.50	12.0	0.0210	8.80
− 0.45	13.1	0.0223	8.69
− 0.40	14.2	0.0235	8.58
− 0.35	15.4	0.0246	8.46
− 0.30	16.7	0.0256	8.33
− 0.25	18.0	0.0265	8.20
− 0.20	19.3	0.0273	8.07
− 0.15	20.7	0.0279	7.93
− 0.10	22.1	0.0284	7.79
− 0.05	23.6	0.0287	7.64
− 0.00	25.0	0.0288	7.50
+ 0.05	26.4	0.0287	7.36
+ 0.10	27.9	0.0284	7.21
+ 0.15	29.3	0.0279	7.07
+ 0.20	30.7	0.0273	6.93
+ 0.25	32.0	0.0265	6.80
+ 0.30	33.3	0.0256	6.67
+ 0.35	34.6	0.0246	6.54
+ 0.40	35.8	0.0235	6.42
+ 0.45	36.9	0.0223	6.31
+ 0.50	38.0	0.0210	6.20
+ 0.55	39.0	0.0197	6.10
+ 0.60	40.0	0.0185	6.00
+ 0.65	40.9	0.0172	5.91
+ 0.70	41.7	0.0160	5.83
+ 0.75	42.5	0.0148	5.75
+ 0.80	43.2	0.0136	5.68
+ 0.85	43.8	0.0125	5.62
+ 0.90	44.4	0.0114	5.56
+ 0.95	45.0	0.0104	5.50
+ 1.00	45.5	0.0095	5.45

*Ml of 0.1M NaOH added to 10 ml 0.5M acid. After addition of NaCl, diluted to give 100 ml (final solution) of 0.05M buffer.
† Buffer capacity.
‡ Ml of 1M NaCl added to give final ionic strength of 0.1.

Table 10.3 *Compositions of 0.05M monoacidic base buffers, adjusted by adding 0.1M HCl, but of varying ionic strength.*

	Titre*	I	β†
pH = pK_a' − 1.00	46.2	0.046	0.010
− 0.95	45.8	0.046	0.010
− 0.90	45.3	0.045	0.011
− 0.85	44.8	0.045	0.012
− 0.80	44.2	0.044	0.014
− 0.75	43.6	0.044	0.015
− 0.70	42.9	0.043	0.016
− 0.65	42.2	0.042	0.017
− 0.60	41.3	0.041	0.018
− 0.55	40.5	0.040	0.020
− 0.50	39.5	0.040	0.021
− 0.45	38.5	0.039	0.022
− 0.40	37.5	0.037	0.023
− 0.35	36.3	0.036	0.025
− 0.30	35.1	0.035	0.026
− 0.25	33.9	0.034	0.027
− 0.20	32.6	0.033	0.027
− 0.15	31.2	0.031	0.028
− 0.10	29.8	0.030	0.028
− 0.05	28.4	0.028	0.029
− 0.00	26.9	0.027	0.029
+ 0.05	25.4	0.025	0.029
+ 0.10	23.9	0.024	0.028
+ 0.15	22.4	0.022	0.028
+ 0.20	21.0	0.021	0.027
+ 0.25	19.5	0.020	0.027
+ 0.30	18.1	0.018	0.026
+ 0.35	16.8	0.017	0.025
+ 0.40	15.5	0.015	0.023
+ 0.45	14.2	0.014	0.022
+ 0.50	13.1	0.013	0.021
+ 0.55	11.9	0.012	0.020
+ 0.60	10.9	0.011	0.018
+ 0.65	9.9	0.010	0.017
+ 0.70	9.0	0.009	0.016
+ 0.75	8.1	0.008	0.015
+ 0.80	7.4	0.007	0.014
+ 0.85	6.6	0.007	0.012
+ 0.90	6.0	0.006	0.011
+ 0.95	5.4	0.005	0.010
+ 1.00	4.8	0.005	0.010

*Ml of 0.1M HCl added to 10 ml 0.5M base. Diluted to give final volume of 100 ml
†Buffer capacity

Table 10.4 *Compositions of 0.05M monobasic acid buffers, adjusted by adding 0.1M NaOH, but of varying ionic strength.*

	Titre*	I	β†
pH = pK_a' − 1.00	4.9	0.005	0.010
− 0.95	5.4	0.005	0.010
− 0.90	6.0	0.006	0.011
− 0.85	6.7	0.007	0.012
− 0.80	7.4	0.007	0.014
− 0.75	8.2	0.008	0.015
−0.70	9.0	0.009	0.016
− 0.65	9.9	0.010	0.017
− 0.60	10.9	0.011	0.018
− 0.55	11.9	0.012	0.020
− 0.50	13.1	0.013	0.021
− 0.45	14.2	0.014	0.022
− 0.40	15.5	0.015	0.023
− 0.35	16.8	0.017	0.025
− 0.30	18.2	0.018	0.026
− 0.25	19.6	0.020	0.027
− 0.20	21.0	0.021	0.027
− 0.15	22.5	0.022	0.028
− 0.10	23.9	0.024	0.028
− 0.05	25.4	0.025	0.029
− 0.00	26.9	0.027	0.029
+ 0.05	28.7	0.029	0.029
+ 0.10	30.1	0.030	0.028
+ 0.15	31.5	0.032	0.028
+ 0.20	32.9	0.033	0.027
+ 0.25	34.2	0.034	0.027
+ 0.30	35.5	0.035	0.026
+ 0.35	36.7	0.037	0.025
+ 0.40	37.8	0.038	0.023
+ 0.45	38.9	0.039	0.022
+ 0.50	39.8	0.040	0.021
+ 0.55	40.8	0.041	0.020
+ 0.60	41.6	0.042	0.018
+ 0.65	42.4	0.042	0.017
+ 0.70	43.1	0.043	0.016
+ 0.75	43.8	0.044	0.015
+ 0.80	44.4	0.044	0.014
+ 0.85	45.0	0.045	0.012
+ 0.90	45.5	0.045	0.011
+ 0.95	45.9	0.046	0.010
+ 1.00	46.3	0.046	0.010

*Ml of 0.1M NaOH added to 10 ml 0.5M acid. Diluted to give final volume of 100 ml
†Buffer capacity

Appendix II:

Composition-pH tables of some commonly used buffers.

Table 10.5 *HCl, KCl buffer (25°C)**

Contains 25 ml 0.2M KCl and x ml 0.2M HCl, diluted to 100 ml

pH	x	β	pH	x	β
1.00	67.0	0.31	1.70	13.0	0.060
1.10	52.8	0.24	1.80	10.2	0.049
1.20	42.5	0.19	1.90	8.1	0.037
1.30	33.6	0.16	2.00	6.5	0.030
1.40	26.6	0.13	2.10	5.1	0.026
1.50	20.7	0.10	2.20	3.9	0.022
1.60	16.2	0.077			

*Bower and Bates (1955).

Table 10.6 *HCl, KCl buffer (25°), I = 0.1**

Contains x ml 0.2M HCl, (50-x) ml 0.2M KCl, diluted to 100 ml

pH	x	β	pH	x	β
1.11	50.0	0.23	1.70	12.8	0.059
1.20	40.7	0.19	1.80	10.2	0.047
1.30	32.3	0.15	1.90	8.1	0.037
1.40	25.7	0.12	2.00	6.5	0.030
1.50	20.1	0.093	2.10	5.2	0.024
1.60	16.0	0.074	2.20	4.2	0.019

*Bower and Bates (1955)

Table 10.7 p-*Toluenesulphonic acid, sodium* p-*toluenesulphonate buffer (25°C)**

Contains x ml 0.2M p-toluenesulphonic acid monohydrate (38.024 $g\,l^{-1}$) and (50-x) ml of 0.2M sodium p-toluenesulphonate (42.40 $g\,l^{-1}$) diluted to 100 ml

pH	x	pH	x	pH	x
1.2	42.0	1.5	18.9	1.8	10.6
1.3	31.6	1.6	15.35	1.9	8.7
1.4	24.8	1.7	12.6	2.0	6.9

*German and Vogel (1937)

Table 10.8 *D(+)-Tartaric acid, NaOH buffer (20°C, 50°C)**

Contains x ml 2M NaOH, 10 ml 2M tartaric acid, diluted to 100 ml

x	pH(20°C)	pH(50°C)	x	pH(20°C)	pH(50°C)
1.0	2.08	2.05	11.0	3.46	3.01
2.0	2.29	2.23	12.0	3.58	3.13
3.0	2.47	2.32	13.0	3.73	3.25
4.0	2.63	2.41	14.0	3.87	3.39
5.0	2.76	2.50	15.0	3.99	3.59
6.0	2.88	2.59	16.0	4.10	3.76
7.0	3.01	2.74	17.0	4.28	4.00
8.0	3.15	2.84	18.0	4.49	4.27
9.0	3.23	2.91	19.0	4.80	4.65
10.0	3.36	3.00	19.5	5.12	5.03

*Gottschalk (1961)

Table 10.9   *Citric acid, NaOH buffer (20°C, 50°C)*†*

Contains x ml 2M NaOH, 10 ml 2M citric acid, diluted to 100 ml

x	pH(20°C)	pH(50°C)	x	pH(20°C)	pH(50°C)
1.0	2.15	2.05	16.0	4.37	4.27
2.0	2.39	2.29	17.0	4.50	4.37
3.0	2.58	2.49	18.0	4.62	4.47
4.0	2.75	2.66	19.0	4.74	4.61
5.0	2.89	2.82	20.0	4.87	4.75
6.0	3.04	2.99	21.0	4.98	4.80
7.0	3.18	3.04	22.0	5.11	4.94
8.0	3.32	3.20	23.0	5.21	5.09
9.0	3.46	3.35	24.0	5.34	5.22
10.0	3.59	3.49	25.0	5.49	5.43
11.0	3.75	3.62	26.0	5.63	5.54
12.0	3.90	3.77	27.0	5.80	5.68
13.0	4.03	3.89	28.0	6.02	5.85
14.0	4.14	4.01	29.0	6.33	6.10
15.0	4.27	4.15	29.5	6.51	6.24

*Gottschalk (1961)
†For citric acid buffers from 12.5°C to 91°C, see Britton and Welford (1937).

Table 10.10   *KH Phthalate, HCl buffer (25°C)**

Contains 50 ml 0.1M KII phthalate and x ml 0.1M HCl, diluted to 100 ml

pH	x	β	pH	x	β
2.20	49.5	0.036	3.20	15.7	0.030
2.30	45.8	0.036	3.30	12.9	0.026
2.40	42.2	0.035	3.40	10.4	0.023
2.50	38.8	0.034	3.50	8.2	0.020
2.60	35.4	0.033	3.60	6.3	0.018
2.70	32.1	0.032	3.70	4.5	0.017
2.80	28.9	0.032	3.80	2.9	0.015
2.90	25.7	0.033	3.90	1.4	0.014
3.00	22.3	0.034	4.00	0.1	0.014
3.10	18.8	0.033			

*Bower and Bates (1955)

Table 10.11   *Glycine, HCl buffer (25°C)**

Contains 50 ml 0.1M glycine and x ml 0.2M HCl, diluted to 100 ml

pH	x	pH	x
2.2	22.0	3.0	5.7
2.4	16.2	3.2	4.1
2.6	12.1	3.4	3.2
2.8	8.4	3.6	2.5

*Gomori (1955)

Table 10.12   *trans-Aconitic acid, NaOH buffer (23°C)**

Contains 25 ml 0.2M aconitic acid (34.82 g l^{-1}) and x ml 0.2M NaOH, diluted to 100 ml

pH	x	pH	x
2.5	7.5	4.2	39.8
2.6	9.0	4.3	41.5
2.7	10.5	4.4	43.3
2.8	12.3	4.5	45.0
2.9	14.0	4.6	46.8
3.0	16.0	4.7	48.5
3.1	18.0	4.8	50.0
3.2	20.0	4.9	51.5
3.3	22.0	5.0	52.8
3.4	24.0	5.1	54.0
3.5	26.0	5.2	55.3
3.6	26.0	5.3	56.5
3.7	30.0	5.4	58.0
3.8	32.0	5.5	59.5
3.9	34.0	5.6	61.3
4.0	36.0	5.7	63.0
4.1	38.0		

*Gomori (1955)

Table 10.13 *Formic acid, sodium formate buffer (0°C, 25°C)**

Contains x ml of M formic acid and y ml M NaOH, diluted to 100 ml

	$I = 0.05$			$I = 0.1$			$I = 0.2$		
	0°C	25°C		0°C	25°C		0°C	25°C	
pH	x	x	y	x	x	y	x	x	y
2.6	74.3	68.4	5.0	—	—	—	—	—	—
2.8	47.8	44.2	5.0	45.9†	84.4	10.0	—	—	—
3.0	31.7	29.4	5.0	61.3	56.6	10.0	—	—	—
3.2	21.7	20.3	5.0	42.2	39.3	10.0	41.6†	77.3	20.0
3.4	15.5	14.6	5.0	30.3	28.4	10.0	59.8	56.1	20.0
3.6	11.6	11.0	5.0	22.8	21.6	10.0	45.1	42.8	20.0
3.8	9.1	8.8	5.0	18.0	17.3	10.0	35.8	34.3	20.0
4.0	7.6	7.4	5.0	15.1	14.6	10.0	30.0	29.1	20.0
4.0	6.6	6.5	5.0	13.2	12.9	10.0	26.3	25.7	20.0
4.4	6.0	6.0	5.0	12.0	11.8	10.0	24.0	23.6	20.0
4.6	5.7	5.6	5.0	11.3	11.1	10.0	22.5	22.3	20.0
4.8	5.4	5.4	5.0	10.8	10.7	10.0	21.6	21.4	20.0

*Long (1961)   †2M solution

Table 10.14 *Citric acid, sodium citrate buffer (23°C)**

Contains x ml 0.1M citric acid (21.01 g $C_6H_8O_7 \cdot H_2O\ l^{-1}$) and (50-$x$) ml 0.1M Na_3 citrate (29.41 g $C_6H_5O_7Na_3 \cdot 2H_2O\ l^{-1}$), diluted to 100 ml

pH	x	pH	x
3.0	46.5	4.8	23.0
3.2	43.7	5.0	20.5
3.4	40.0	5.2	18.0
3.6	37.0	5.4	16.0
3.8	35.0	5.6	13.7
4.0	33.0	5.8	11.8
4.2	31.5	6.0	9.5
4.4	28.0	6.2	7.2
4.6	25.5		

*Gomori (1955)
For disodium hydrogen citrate, NaOH, HCl buffers covering the pH range 2.2−6.8, see Sörensen (1909, 1912)

Table 10.15 *3,3-Dimethylglutaric acid, NaOH buffer (21°C)**

Contains 100 ml 0.1M dimethylglutaric acid (16.02 g l^{-1}) and x ml 0.2M NaOH, diluted to 1 litre.

pH	x	pH	x
3.2	8.3	5.6	55.8
3.4	14.7	5.8	59.7
3.6	22.0	6.0	65.0
3.8	27.4	6.2	70.5
4.0	33.3	6.4	75.5
4.2	36.8	6.6	84.7
4.4	39.8	6.8	88.0
4.6	41.7	7.0	90.4
4.8	43.9	7.2	92.1
5.0	46.2	7.4	93.2
5.2	49.0	7.6	94.0
5.4	52.0		

*Stafford *et al.*, (1955)

Table 10.16 *3,3-Dimethylglutaric acid, NaOH buffer (21°C), 0.1M NaCl*

Contains 5.845 g NaCl, 100 ml 0.1M dimethylglutaric acid (16.02 g l^{-1}), and x ml 0.2M NaOH, diluted to 1 litre.

pH	x	pH	x
3.2	15.7	5.6	64.6
3.4	22.1	5.8	69.7
3.6	27.9	6.0	74.6
3.8	33.3	6.2	79.5
4.0	37.4	6.4	84.2
4.2	40.9	6.6	88.1
4.4	43.6	6.8	91.5
4.6	46.2	7.0	93.6
4.8	48.9	7.2	95.1
5.0	52.0	7.4	96.0
5.2	54.3	7.6	97.0
5.4	59.2		

*Stafford *et al.*, (1955)

Table 10.17 *Phenylacetic acid, sodium phenylacetate buffer (25°C)**

Contains x ml 0.01M sodium phenylacetate (1.5806 g l⁻¹) and (50-x) ml 0.01M phenylacetic acid (1.3606 g l⁻¹), diluted to 100 ml

pH	x	pH	x	pH	x
3.40	1.3	4.10	19.2	4.70	35.6
3.50	3.4	4.20	22.0	4.80	37.9
3.60	5.6	4.30	24.9	4.90	39.8
3.70	7.8	4.40	28.0	5.00	41.4
3.80	10.1	4.50	30.6	5.10	42.8
3.90	12.8	4.60	33.1	5.20	44.3
4.00	15.6				

*German & Vogel (1937)

Table 10.18 *Sodium acetate, acetic acid buffer (23°C)**

Contains x ml 0.2M acetic acid and (50-x) ml 0.2M NaOAc, diluted to 100 ml

pH	x	pH	x
3.6	46.3	4.8	20.0
3.8	44.0	5.0	14.8
4.0	41.0	5.2	10.5
4.2	36.8	5.4	8.8
4.4	30.5	5.6	4.8
4.6	25.5		

*Gomori (a955)

Table 10.18 (alternative) *Sodium acetate, acetic acid buffer (20°C)**

Contains x ml 2M NaOAc and $(10-x)$ ml of 2M acetic acid, diluted to 100 ml

pH	x	pH	x	pH	x
3.40	0.50	4.40	3.6	5.20	7.8
3.70	1.0	4.50	4.15	5.30	8.15
3.80	1.25	4.60	4.7	5.40	8.45
3.90	1.55	4.70	5.3	5.50	8.75
4.00	1.85	4.80	5.85	5.60	9.0
4.10	2.2	4.90	6.4	5.90	9.5
4.20	2.6	5.00	6.95		
4.30	3.05	5.10	7.4		

*Based on Gottschalk (1959).

Table 10.19 *Succinic acid, NaOH buffer (23°C)**

Contains 25 ml 0.2M succinic acid (23.62 g l^{-1}) and x ml 0.2M NaOH, diluted to 100 ml

pH	x	pH	x
3.8	7.5	5.0	26.7
4.0	10.0	5.2	30.3
4.2	13.3	5.4	34.2
4.4	16.7	5.6	37.5
4.6	20.0	5.8	40.7
4.8	23.5	6.0	43.5

*Gomori (1955)

Table 10.20   *KH Phthalate, NaOH buffer (25°C)**

Contains 50 ml 0.1M KH Phthalate (20.42 g l^{-1}) and x ml 0.1M NaOH, diluted to 100 ml

pH	x	β
4.10	1.3	0.016
4.20	3.0	0.017
4.30	4.7	0.018
4.40	6.6	0.020
4.50	8.7	0.022
4.60	11.1	0.025
4.70	13.6	0.027
4.80	16.5	0.029
4.90	19.4	0.030
5.00	22.6	0.031
5.10	25.5	0.031
5.20	28.8	0.030
5.30	31.6	0.026
5.40	34.1	0.025
5.50	36.6	0.023
5.60	38.8	0.020
5.70	40.6	0.017
5.80	42.3	0.015
5.90	43.7	0.013

*Bower and Bates (1955)

Table 10.21   *Sodium cacodylate, HCl buffer (15°C)**

Contains 50 ml 0.1M Na(CH$_3$)$_2$ AsO$_2$·3H$_2$O (21.4 g l^{-1}) and x ml 0.1M HCl, diluted to 100 ml

pH	x	pH	x
5.0	46.75	6.4	18.75
5.2	45.05	6.6	13.3
5.4	42.6	6.8	9.3
5.6	39.2	7.0	6.3
5.8	34.8	7.2	4.15
6.0	29.55	7.4	2.7
6.2	23.85		

*Plumel (1948)

Table 10.22   *Sodium hydrogen maleate, NaOH buffer (25°C)**

Contains 25 ml 0.2M sodium hydrogen maleate[†] and x ml 0.1M NaOH, diluted to 100 ml

pH	x	pH	x
5.2	7.2	6.2	33.0
5.4	10.5	6.4	38.0
5.6	15.3	6.6	41.6
5.8	20.8	6.8	44.4
6.0	26.9		

*Temple (1929).
[†]Prepared by dissolving 23.2 g maleic acid in water, adding 200 ml 1M NaOH and diluting to 1 litre.
    For a buffer table at 18°C of mixtures of 0.1M NaH maleate and 0.1M Na_2 maleate, see Smits (1947).

Table 10.23   *Maleic acid, Tris, NaOH buffer (23°C)**

Contains 25 ml 0.2M Tris[†] and 0.2M maleic acid,[†] and x ml 0.2M NaOH, diluted to 100 ml

pH	x	pH	x
5.2	3.5	7.0	24.0
5.4	5.4	7.2	25.5
5.6	7.75	7.4	27.0
5.8	10.25	7.6	29.0
6.0	13.0	7.8	31.65
6.2	15.75	8.0	34.5
6.4	18.5	8.2	37.5
6.6	21.25	8.4	40.5
6.8	22.9	8.6	43.3

*Gomori (1955).
[†]Prepared by dissolving 23.2 g maleic acid and 24.2 g tris(hydroxymethyl)aminomethane in 1 litre.

Table 10.24   *Phosphate buffer (25°C)**

Contains x ml 0.2M Na_2HPO_4[†] and (50-x) ml 0.2M NaH_2PO_4[‡], diluted to 100 ml

pH	x	pH	x
5.8	4.0	7.0	30.5
6.0	6.15	7.2	36.0
6.2	9.25	7.4	40.5
6.4	13.25	7.6	43.5
6.6	18.75	7.8	45.75
6.8	24.5	8.0	47.35

*Gomori (1955).
[†] 28.39 g Na_2HPO_4, 35.61 g $Na_2HPO_4 \cdot 2H_2O$ or 71.64 g $Na_2HPO_4 \cdot 12H_2O$ 1^{-1}
[‡] 27.60 g $NaH_2PO_4 \cdot H_2O$ or 31.21 g $NaH_2PO_4 \cdot 2H_2O$ 1^{-1}

Table 10.25 *KH_2PO_4, NaOH buffer (25°C)**

Contains 50 ml 0.1M KH_2PO_4 (13.60 g l^{-1}) and
x ml 0.1M NaOH, diluted to 100 ml

pH	x	β
5.80	3.6	—
5.90	4.6	0.010
6.00	5.6	0.011
6.10	6.8	0.012
6.20	8.1	0.015
6.30	9.7	0.017
6.40	11.6	0.021
6.50	13.9	0.024
6.60	16.4	0.027
6.70	19.3	0.030
6.80	22.4	0.033
6.90	25.9	0.033
7.00	29.1	0.031
7.10	32.1	0.028
7.20	34.7	0.025
7.30	37.0	0.022
7.40	39.1	0.020
7.50	40.9	0.016
7.60	42.4	0.013
7.70	43.5	0.011
7.80	44.5	0.009
7.90	45.3	0.008
8.00	46.1	—

*Bower & Bates (1955).

Table 10.26 *Imidazole, HCl buffer (25°C)*

Contains 25 ml 0.2M imidazole (13.62 g l^{-1}) and
x ml 0.2M HCl, diluted to 100 ml

pH	x	pH	x
6.2	21.45	7.2	9.3
6.4	19.9	7.4	6.8
6.6	17.75	7.6	4.65
6.8	15.2	7.8	3.0
7.0	12.15		

*Mertz & Owen (1940).

Table 10.27   *2,4,6-Trimethylpyridine, HCl buffer (23°C and 37°C)**

Contains   25 ml   0.2M   2,4,6-trimethylpyridine
(24.24 g l^{-1})   and x ml   0.2M   HCl, diluted   to
100 ml

pH, 23°C	pH, 37°C	x
6.45	6.37	22.5
6.62	6.54	21.25
6.80	6.72	20.0
6.92	6.84	18.75
7.03	6.95	17.5
7.13	7.05	16.25
7.22	7.14	15.0
7.30	7.23	13.75
7.40	7.32	12.5
7.50	7.40	11.25
7.57	7.50	10.0
7.67	7.60	8.75
7.77	7.70	7.5
7.88	7.80	6.25
8.00	7.94	5.0
8.18	8.10	3.75
8.35	8.28	2.5

*Gomori (1946).

Table 10.28 *Triethanolamine   hydrochloride,
NaOH buffer (20°C)**

Contains 50 ml 0.1M triethanolamine hydrochloride (18.57 g l^{-1}) and x ml 0.1M NaOH,
diluted to 100 ml

pH	x	pH	x
6.8	3.3	7.9	23.8
6.9	4.1	8.0	26.5
7.0	5.1	8.1	29.4
7.1	6.2	8.2	32.1
7.2	7.6	8.3	34.7
7.3	9.2	8.4	37.0
7.4	11.1	8.5	39.1
7.5	13.2	8.6	40.9
7.6	15.5	8.7	42.5
7.7	18.1	8.8	43.9
7.8	20.8		

*Computed using Davies' equation.

Table 10.29 *Sodium 5,5-diethylbarbiturate,
HCl buffer (18°C)**

Contains 100 ml 0.04M sodium diethylbarbiturate (8.25 g l^{-1}) to which is then added x ml
0.2M HCl.   HCl.

pH	x	pH	x
6.8	18.4	8.2	7.21
7.0	17.8	8.4	5.21
7.2	16.7	8.6	3.82
7.4	15.3	8.8	2.52
7.6	13.4	9.0	1.65
7.8	11.5	9.2	1.13
8.0	9.39		

*Britton & Robinson (1931a).
For buffers of x ml 0.1M sodium diethylbarbiturate and $(100-x)$ ml of 0.1M HCl, see
Michaelis (1930), and for the effect of temperature on diethylbarbiturate buffers used in electrophoresis, see Strickland and Anderson (1966).

Table 10.30   N-Ethylmorpholine, HCl buffer
(20°C)*

Contains   50 ml   0.2M   *N*-ethylmorpholine
(23.03 g l^{-1}) and *x* ml 0.2M HCl, diluted to 100 ml

pH	x	pH	x
6.8	46.2	7.8	26.9
7.0	44.2	8.0	20.9
7.2	41.4	8.2	15.4
7.4	37.5	8.4	10.8
7.6	32.6	8.6	7.3

*Computed using Davies' equation.

Table 10.31   *Sodium pyrophosphate, HCl buffer (25°C)**

Contains *x* ml 0.1M HCl and 50 ml 0.1M $Na_4P_2O_7 \cdot 10H_2O$
(44.61 g l^{-1})

pH	x	pH	x	pH	x
6.90	50.0	8.10	26.9	8.95	8.9
7.63	40.9	8.30	21.4	–	–
7.95	33.3	8.67	12.5	–	–

*Israel (1949).

Table 10.32 *Tris(hydroxymethyl)aminomethane, HCl buffer (25°C)*†*

Contains 50 ml 0.1M Tris (12.114 g l^{-1}) and x ml 0.1M HCl, diluted to 100 ml

pH	x	β
7.00	46.6	0.0085
7.10	45.7	0.010
7.20	44.7	0.012
7.30	43.4	0.013
7.40	42.0	0.015
7.50	40.3	0.017
7.60	38.5	0.018
7.70	36.6	0.020
7.80	34.5	0.023
7.90	32.0	0.027
8.00	29.2	0.029
8.10	26.2	0.031
8.20	22.9	0.031
8.30	19.9	0.029
8.40	17.2	0.026
8.50	14.7	0.024
8.60	12.4	0.022
8.70	10.3	0.020
8.80	8.5	0.016
8.90	7.0	0.014
9.00	5.7	0.011

*Bates & Bower (1956).
†$d(pH)/dT \approx -0.028$.

Table 10.33 *Bicine, NaOH buffer (20°C, 37°C)**

Contains x ml 0.1M NaOH added to 40 ml 0.1M Bicine (16.32 g l^{-1})

pH (20°C)	x	pH (20°C)	pH (37°C)	x
7.0	2.0	7.95	–	11.0
7.15	3.0	8.0	7.8	12.0
7.35	4.0	8.1	7.95	15.0
7.5	5.0	8.2	8.05	17.0
7.6	6.0	8.3	8.1	19.0
7.65	7.0	8.35	8.15	20.0
7.8	9.0	8.45	8.30	23.0
7.85	10.0			

*Remizov (1960).

Table 10.34 *2-Amino-2-methylpropane-1,3-diol, HCl buffer (23°C and 37°C)**

Contains 25 ml 0.2M 2-amino-2-methylpropane-1,3-diol (21.03 g l^{-1}) and x ml 0.2M HCl, diluted to 100 ml

pH (23°C)	pH (37°C)	x
7.83	7.72	22.5
8.00	7.90	21.25
8.18	8.07	20.0
8.30	8.20	18.75
8.40	8.30	17.5
8.50	8.40	16.25
8.60	8.50	15.0
8.70	8.58	13.75
8.78	8.67	12.5
8.87	8.76	11.25
8.96	8.85	10.0
9.05	8.94	8.75
9.15	9.03	7.5
9.26	9.15	6.25
9.38	9.27	5.0
9.56	9.45	3.75
9.72	9.62	2.5

*Gomori (1946).

Table 10.35 *Diethanolamine, HCl buffer (25°C)**

Contains 25 ml 0.2M diethanolamine (21.02 g l^{-1}) and x ml 0.2M HCl, diluted to 100 ml

pH	x	I	β
7.80	23.40	0.0468	0.0082
7.90	23.01	0.0460	0.0099
8.00	22.54	0.0451	0.0118
8.10	21.98	0.0440	0.0141
8.20	21.30	0.0426	0.0165
8.30	20.51	0.0410	0.0190
8.40	19.58	0.0392	0.0215
8.50	18.52	0.0370	0.0239
8.60	17.32	0.0346	0.0260
8.70	16.01	0.0320	0.0276
8.80	14.61	0.0292	0.0285
8.90	13.15	0.0263	0.0288
9.00	11.66	0.0233	0.0282
9.10	10.20	0.0204	0.0270
9.20	8.80	0.0176	0.0252
9.30	7.49	0.0150	0.0230
9.40	6.30	0.0126	0.0205
9.50	5.24	0.0105	0.0180
9.60	4.32	0.0086	0.0155
9.70	3.54	0.0071	0.0131
9.80	2.87	0.0057	0.0110
9.90	2.32	0.0046	0.0092

*Computed using Davies' equation. Computed pH values agree within 0.05 pH units with values quoted by Dawson *et al.*, (1969).

Table 10.36 *Potassium p-phenolsulphonate, NaOH Buffer (0°C, 25°C)**

Contains x ml M potassium p-phenolsulphonate and y ml M NaOH, diluted to 1 litre

	$I = 0.05$		$I = 0.05$		$I = 0.1$		$I = 0.1$	
	0°C		25°C		0°C		25°C	
pH	x	y	x	y	x	y	x	y
7.8	–	–	42.7	3.6	–	–	84.2	7.9
8.0	44.1	3.0	39.8	5.1	87.1	6.4	78.0	11.0
8.2	41.5	4.3	36.3	6.9	81.7	9.2	70.7	14.6
8.4	38.3	5.9	32.5	8.8	75.0	12.5	63.1	18.4
8.6	34.6	7.7	28.8	10.6	67.5	16.3	55.8	22.1
8.8	30.8	9.6	25.5	12.3	59.9	20.1	49.5	25.2
9.0	27.2	11.4	22.8	13.6	53.0	23.5	44.6	27.7
9.2	24.2	12.9	20.8	14.6	47.2	26.4	40.9	29.6
9.4	21.8	14.1	19.4	15.3	42.8	28.6	38.3	30.6
9.6	20.1	14.9	18.5	15.8	39.7	30.2	36.6	31.7
9.8	18.9	15.5	17.8	16.1	37.5	31.3	35.4	32.3
10.0	18.1	15.9	17.4	16.3	36.0	32.0	34.7	32.7

*Long (1961).

Table 10.37 *Boric acid, NaOH buffer (25°C)**

Contains 50 ml 0.1M boric acid† and 0.1M KCl†
and x ml 0.1M NaOH, diluted to 100 ml

pH	x	β
8.00	3.9	—
8.10	4.9	0.010
8.20	6.0	0.011
8.30	7.2	0.013
8.40	8.6	0.015
8.50	10.1	0.016
8.60	11.8	0.018
8.70	13.7	0.020
8.80	15.8	0.022
8.90	18.1	0.025
9.00	20.8	0.027
9.10	23.6	0.028
9.20	26.4	0.029
9.30	29.3	0.028
9.40	32.1	0.027
9.50	34.6	0.024
9.60	36.9	0.022
9.70	38.9	0.019
9.80	40.6	0.016
9.90	42.2	0.015
10.00	43.7	0.014
10.10	45.0	0.013
10.20	46.2	—

*Bower and Bates (1955).
†Prepared by dissolving 6.184 g H_3BO_3 and
7.455 g KCl in 1 litre. For boric acid, NaOH
buffers from 12.5°C to 91°C see Britton and
Welford (1937).

Table 10.37a   *Sodium borate, HCl buffer (25°C)*†*

Contains 50 ml 0.025M $Na_2B_4O_7 \cdot 10H_2O$ (9.525 g l^{-1}), and x ml 0.1M HCl, diluted to 100 ml

pH	x	β
8.00	20.5	–
8.10	19.7	0.009
8.20	18.8	0.010
8.30	17.7	0.011
8.40	16.6	0.012
8.50	15.2	0.015
8.60	13.5	0.018
8.70	11.6	0.020
8.80	9.4	0.023
8.90	7.1	0.024
9.00	4.6	0.026
9.10	2.0	–

*Bates & Bower (1956).
†$d(pH)/dT \approx -0.008$.

Table 10.38   *Ammonia/ammonium chloride buffer (20°C)**

Contains x ml 2M NH_3 and $(10 - x)$ ml 2M NH_4Cl, diluted to 100 ml

pH	x	pH	x	pH	x
8.25	0.50	9.21	3.0	10.18	8.0
8.61	1.0	9.58	5.0	10.51	9.0
8.96	2.0	9.94	7.0	10.82	9.5

*Gottschalk (1959).

Table 10.39 *Glycine, NaOH buffer (25°C)**

Contains 25 ml 0.2M glycine (15.01 g l^{-1}) and x ml 0.2M NaOH, diluted to 100 ml

pH	x	pH	x
8.6	2.0	9.6	11.2
8.8	3.0	9.8	13.6
9.0	4.4	10.0	16.0
9.2	6.0	10.4	19.3
9.4	8.4	10.6	22.75

*Gomori (1955).

Table 10.40 *Na_2CO_3, $NaHCO_3$ buffer (20°C and 37°C)**

Contains x ml 0.1M $Na_2CO_3 \cdot 10H_2O$ (28.62 g l^{-1}) and $(100 - x)$ ml 0.1M $NaHCO_3$ (8.40 g l^{-1})

pH (20°C)	pH (37°C)	x
9.16	8.77	10
9.40	9.12	20
9.51	9.40	30
9.78	9.50	40
9.90	9.72	50
10.14	9.90	60
10.28	10.08	70
10.53	10.28	80
10.83	10.57	90

*Delory & King (1945).

Table 10.41   *Sodium borate, NaOH buffer (25°C)*†*

Contains 50 ml 0.025M $Na_2B_4O_7 \cdot 10H_2O$ (9.525 g l^{-1}) and x ml 0.1M NaOH, diluted to 100 ml

pH	x	β
9.20	0.9	—
9.30	3.6	0.027
9.40	6.2	0.026
9.50	8.8	0.025
9.60	11.1	0.022
9.70	13.1	0.020
9.80	15.0	0.018
9.90	16.7	0.016
10.00	18.3	0.014
10.10	19.5	0.011
10.20	20.5	0.009
10.30	21.3	0.008
10.40	22.1	0.007
10.50	22.7	0.006
10.60	23.3	0.005
10.70	23.8	0.004
10.80	24.25	—

*Bates & Bower (1956).
†$d(pH)/dT \approx -0.008$.

Table 10.42   *NaHCO₃, NaOH buffer (25°C)*†*

Contains 50 ml 0.05M NaHCO₃ (4.20 g l⁻¹) and
x ml 0.1M NaOH, diluted to 100 ml

pH	x	β
9.60	5.0	—
9.70	6.2	0.013
9.80	7.6	0.014
9.90	9.1	0.015
10.00	10.7	0.016
10.10	12.2	0.016
10.20	13.8	0.015
10.30	15.2	0.014
10.40	16.5	0.013
10.50	17.8	0.013
10.60	19.1	0.012
10.70	20.2	0.010
10.80	21.2	0.009
10.90	22.0	0.008
11.00	22.7	—

*Bates & Bower (1956).
†d(pH)/d$T \approx -0.009$.

Table 10.43   *Na₂HPO₄, NaOH buffer (25°C)*†*

Contains 50 ml 0.05M Na₂HPO₄ (7.10 g l⁻¹) and
x ml 0.1M NaOH, diluted to 100 ml

pH	x	β
10.90	3.3	—
11.00	4.1	0.009
11.10	5.1	0.011
11.20	6.3	0.012
11.30	7.6	0.014
11.40	9.1	0.017
11.50	11.1	0.022
11.60	13.5	0.026
11.70	16.2	0.030
11.80	19.4	0.034
11.90	23.0	0.037
12.00	26.9	—

*Bates & Bower (1956).
†d(pH)/d$T \approx -0.025$.

Table 10.44 *NaOH, KCl buffer (25°C)*†*

Contains 25 ml 0.2M KCl (14.91 g l⁻¹) and
x ml 0.2M NaOH, diluted to 100 ml

pH	x	β
12.00	6.0	0.028
12.10	8.0	0.042
12.20	10.2	0.048
12.30	12.8	0.060
12.40	16.2	0.076
12.50	20.4	0.094
12.60	25.6	0.12
12.70	32.2	0.16
12.80	41.2	0.21
12.90	53.0	0.25
13.00	66.0	0.30

*Bates & Bower (1956).
†d(pH)/d$T \approx -0.033$.

Table 10.45   *Wide range buffer (pH 2.6–8.0) (21°C)**

Contains x ml 0.1M citric acid† and $(100 - x)$ ml 0.2M Na_2HPO_4 ‡.

pH	x	I §	β
2.2	98.0	–	–
2.4	94.9	–	–
2.6	90.3	0.044	0.061
2.8	85.8	0.062	0.066
3.0	81.1	0.081	0.066
3.2	76.6	0.100	0.061
3.4	72.4	0.12	0.056
3.6	68.7	0.135	0.050
3.8	65.2	0.15	0.047
4.0	61.9	0.17	0.046
4.2	59.0	0.19	0.046
4.4	56.3	0.205	0.044
4.6	53.8	0.22	0.042
4.8	51.4	0.24	0.039
5.0	49.0	0.25	0.036
5.2	47.0	0.27	0.036
5.4	44.8	0.28	0.036
5.6	42.6	0.30	0.036
5.8	40.2	0.31	0.036
6.0	37.5	0.32	0.037
6.2	34.6	0.34	0.040
6.4	31.1	0.35	0.048
6.6	27.1	0.36	0.060
6.8	22.8	0.37	0.076
7.0	17.8	0.40	0.092
7.2	13.0	0.43	0.101
7.4	9.4	0.46	0.100
7.6	6.5	0.49	0.088
7.8	4.2	0.52	0.071
8.0	2.8	0.55	0.053

*McIlvaine (1921); Whiting (1966).
†21.01 g $C_6H_8O_7 \cdot H_2O$ l^{-1}.
‡35.61 g $Na_2 HPO_4 \cdot 2H_2O$ or 28.40 g Na_2HPO_4 l^{-1}.
§ To adjust these buffers to a constant ionic strength of 0.5 to 1.0, by adding KCl, see Elving *et al.*, (1956).

Table 10.46   *Wide range buffer (pH 4.4–10.8) (25°C)**

Contains 1 litre 0.01M piperazine dihydrochloride† and 0.01M glycylglycine† and x ml 1M NaOH.

pH	x	pH	x
4.4	0.07	7.8	13.29
4.6	0.54	8.0	14.45
4.8	1.20	8.2	15.71
5.0	2.20	8.4	16.97
5.2	3.20	8.6	18.25
5.4	4.19	8.8	19.50
5.6	5.38	9.0	20.64
5.8	6.51	9.2	21.85
6.0	7.45	9.4	23.13
6.2	8.27	9.6	24.51
6.4	8.99	9.8	25.87
6.6	9.53	10.0	27.17
6.8	9.99	10.2	28.32
7.0	10.45	10.4	29.37
7.2	10.91	10.6	30.42
7.4	11.52	10.8	31.53
7.6	12.31		

*Smith and Smith (1949).
†Made by dissolving 1.591 g piperazine dihydrochloride and 1.321 g glycylglycine in 1 litre.

Table 10.47   *Universal buffer (pH 2.6–12), (18° C)*†*

100 ml of a solution 0.0286M in citric acid (6.008 g l^{-1}), 0.0286M in KH_2PO_4 (3.893 g l^{-1}), 0.0286M in boric acid (1.769 g l^{-1}), and 0.0286M in diethylbarbituric acid (5.266 g l^{-1}), as stock solution, to which is added x ml 0.2M NaOH, followed by dilution to 200 ml‡

pH	x	pH	x	pH	x
2.58	0	5.70	36	8.76	70
2.72	2	5.91	38	8.97	72
2.86	4	6.10	40	9.20	74
3.03	6	6.28	42	9.41	76
3.21	8	6.45	44	9.65	78
3.43	10	6.62	46	9.88	80
3.66	12	6.79	48	10.21	82
3.87	14	6.94	50	10.63	84
4.09	16	7.12	52	11.00	86
4.26	18	7.30	54	11.23	88
4.42	20	7.45	56	11.44	90
4.58	22	7.62	58	11.60	92
4.72	24	7.79	60	11.75	94
4.91	26	7.98	62	11.85	96
5.08	28	8.12	64	11.94	98
5.25	30	8.35	66	12.02	100
5.40	32	8.55	68		
5.57	34				

*Britton and Robinson (1931).

†For pH values of this buffer over the temperature range, 12.5°C–91°C, see Britton and Welford (1937).

‡At pH values greater than 4.5, the dilution step can be omitted without changing the pH.

Table 10.48   *Universal buffer (pH 2.6–12) (18°C)*

Same stock solution as for Table 10.47, to which is added x ml of 0.2M NaOH, without further dilution.

pH	x	pH	x	pH	x
2.6	2.0	5.8	36.5	9.0	72.7
2.8	4.3	6.0	38.9	9.2	74.0
3.0	6.4	6.2	41.2	9.4	75.9
3.2	8.3	6.4	43.5	9.6	77.6
3.4	10.1	6.6	46.0	9.8	79.3
3.6	11.8	6.8	48.3	10.0	80.8
3.8	13.7	7.0	50.6	10.2	82.0
4.0	15.5	7.2	52.9	10.4	82.9
4.2	17.6	7.4	55.8	10.6	83.9
4.4	19.9	7.6	58.6	10.8	84.9
4.6	22.4	7.8	61.7	11.0	86.0
4.8	24.8	8.0	63.7	11.2	87.7
5.0	27.1	8.2	65.6	11.4	89.7
5.2	29.5	8.4	67.5	11.6	92.0
5.4	31.8	8.6	69.3	11.8	95.0
5.6	34.2	8.8	71.0	12.0	99.6

Table 10.49   *Approximate* universal buffer†*

Contains x ml solution 0.2M in boric acid and 0.05M in citric acid, and $(200 - x)$ ml solution 0.1M trisodium orthophosphate. $12H_2O$.

pH	x	pH	x	pH	x
2.0	195	5.5	126	9.0	69
2.5	184	6.0	118	9.5	60
3.0	176	6.5	109	10.0	54
3.5	166	7.0	99	10.5	49
4.0	155	7.5	92	11.0	44
4.5	144	8.0	85	11.5	33
5.0	134	8.5	78	12.0	17

*These buffer solutions are very easily prepared but the pH values are only approximate because of variation in composition of commercially available trisodium phosphate. The table was based on the use of commercial reagent containing approximately 2% NaOH.
†Carmody (1961).

APPENDIX III:

Thermodynamic acid dissociation constants of prospective buffer substances

pK_a (25°C)	Substance	dpK_a/dT
0.66	Trichloroacetic acid	
0.85	Pyrophosphoric acid (pK_1)	
1.27	Oxalic acid (pK_1)	
1.96	Pyrophosphoric acid (pK_2)	
1.96	Sulphuric acid (pK_2)	0.015
2.00	Maleic acid (pK_1)	
2.15	o-Aminobenzoic acid	
2.15	Phosphoric acid (pK_1)	
2.35	Glycine (pK_1)	0.0044
2.71	Alanine (pK_1)	−0.0020
2.80	*trans*-Aconitic acid (pK_1)	
2.88	Chloroacetic acid	0.0023
2.88	Malonic acid (pK_1)	
2.95	Phthalic acid (pK_1)	0.0020
2.96	Diglycollic acid (pK_1)	
2.98	Salicylic acid (pK_1)	
3.03	Fumaric acid (pK_1)	
3.04	D(+)-Tartaric acid (pK_1)	−0.0020
3.13	Citric acid (pK_1)	−0.0024
3.14	Glycylglycine (pK_1)	~0
3.17	Furoic acid	

APPENDIX III: *(Cont.)*

pK_a (25°C)	Substance	dpK_a/dT
3.22	Sulphanilic acid	
3.36	Mandelic acid	
3.40	Malic acid (pK_1)	
3.64	Hippuric acid	
3.70	3,3-Dimethylglutaric acid (pK_1)	0.0076
3.75	Formic acid	~0
3.83	Glycollic acid	~0
3.86	Lactic acid	~0
4.04	Barbituric acid	
4.20	Benzoic acid	0.018
4.21	Succinic acid (pK_1)	−0.0018
4.29	Oxalic acid (pK_2)	0.0038
4.37	D(+)-Tartaric acid (pK_2)	~0
4.38	Fumaric acid (pK_2)	
4.43	Diglycollic acid (pK_2)	
4.46	*trans*-Aconitic acid (pK_2)	
~4.5‡	Tetrakis-(2-hydroxyethyl)ethylenediamine (pK_2)	
4.66	Aniline	
4.76	Acetic acid	0.0002
4.76	Citric acid (pK_2)	−0.0016
4.80	Valeric acid	
4.83	Butyric acid	
4.83	Isobutyric acid	

APPENDIX III: *(Cont.)*

pK$_a$ (25°C)	Substance	dpK$_a$/dT
4.86	Propionic acid	
5.0	Quinoline	
5.13	Malic acid (pK$_2$)	−0.014
5.23	Pyridine	
5.30	*p*-Toluidine	
5.41	Phthalic acid (pK$_2$)	−0.001 (<25°C)
		+0.001 (>25°C)
5.55	Piperazine (pK$_2$)	−0.015
5.64	Succinic acid (pK$_2$)	~0
5.68	Malonic acid (pK$_2$)	
5.83	Uric acid	
5.89‡	Tetraethylethylenediamine (pK$_2$)	
5.96	Histidine (pK$_2$)	
6.03§	2,4,6-Trichlorophenol	
6.15 (20°C)*	MES (2-(*N*-Morpholino)ethanesulphonic acid)	−0.011
6.26	Maleic acid (pK$_2$)	
6.27	Cacodylic acid (Dimethylarsinic acid)	
6.34	3,3-Dimethylglutaric acid (pK$_2$)	0.0060
6.35	Carbonic acid (apparent pK$_1$)	−0.0055
6.39	4-Hydroxymethylimidazole	−0.018
6.40	Citric acid (pK$_3$)	~0
6.46 (20°C)*	Bis-tris [bis-(2-hydroxyethyl)imino]-tris [(hydroxymethyl)methane]	
6.5§	Orthophosphorous acid (pK$_2$′)	

APPENDIX III: (*Cont.*)

pK$_a$ (25°C)	Substance	dpK$_a$/dT
6.50‡	Dimethylaminoethylamine (pK$_2$)	−0.011
6.62 (20°C)*	ADA (N-(2-Acetamido)iminodiacetic acid)	
6.60	Pyrophosphoric acid (pK$_3$)	
6.65 (18°C)†	N,N'-Bis(3-sulphopropyl)ethylenediamine	
6.65	Glycerol-2-phosphoric acid (pK$_2$)	~0
6.80 (20°C)*	PIPES (Piperazine-N,N'-bis(2-ethanesulphonic acid))	−0.0085
6.8 (20°C)*	Bis-tris-propane (1,3-bis[tris(hydroxymethyl)methylamino]propane)	
6.85	Ethylenediamine (pK$_2$)	−0.027
6.88 (20°C)*	ACES (N-(2-Acetamido)-2-aminoethanesulphonic acid)	−0.020
6.95	Imidazole	−0.020
6.98	Arsenic acid (pK$_2$)	−0.001
7.10 (20°C)*	(2-Aminoethyl)trimethylammonium chloride (Cholamine chloride)	−0.027
7.15	p-Nitrophenol	
7.17 (20°C)*	BES (N,N-Bis(2-hydroxyethyl)-2-aminoethanesulphonic acid)	−0.016
7.20 (20°C)*	MOPS (3-(N-Morpholino)propanesulphonic acid)	
7.20	Phosphoric acid (pK$_2$)	−0.0028
7.23	3,6-Endomethylene-1,2,3,6-tetrahydrophthalic acid (pK$_2$)‖	
7.4‡	2,3-Dihydroxypropyl-tris-(hydroxymethyl)methylamine	
7.43	2,4,6-Trimethylpyridine	−0.022
7.50 (20°C)*	TES (N-Tris(hydroxymethyl)methyl-2-aminoethanesulphonic acid)	−0.020
7.52	4-Methylimidazole	−0.022
7.55 (20°C)*	HEPES (N-2-Hydroxyethylpiperazine-N'-2-ethanesulphonic acid)	−0.014
7.67	N-Ethylmorpholine	−0.022

APPENDIX III: (*Cont.*)

pK_a (25°C)	Substance	dpK_a/dT
7.76	Triethanolamine	−0.020
7.83	Mono-tris 2-Hydroxyethylimino-tris(hydroxymethyl)methane	
7.86	Triisopropanolamine	−0.014
7.98	5,5-Diethylbarbituric acid	
8.00 (20°C)*	EPPS (HEPPS) (N-2-Hydroxyethylpiperazinepropanesulphonic acid)	
8.06	Tris (Tris(hydroxymethyl)aminomethane)	−0.028
8.1 (18°C)†	1,4-Bis(3-sulphopropyl)piperazine	
8.15 (20°C)*	Tricine (N-Tris(hydroxymethyl)methylglycine)	−0.021
8.20 (20°C)*	Glycinamide	−0.029
8.25	Glycylglycine (pK_2)	−0.025
8.35 (20°C)*	Bicine (N,N-Bis(2-hydroxyethyl)glycine)	−0.018
8.36	2,5-Dimethylimidazole	−0.025
8.4 (20°C)*	TAPS (Tris(hydroxymethyl)methylaminopropanesulphonic acid)	
~8.5§	Tetrakis-(2-hydroxyethyl)ethylenediamine (pK_1)	
8.52	N-Methyldiethanolamine	
8.6 (18°C)†	1,4-Bis(4-sulphobutyl)piperazine	
8.79	2-Amino-2-methyl-1,3-propanediol	−0.029
8.80	2-Amino-2-ethyl-1,3-propanediol	−0.029
8.81	Ethanolisopropanolamine	−0.024
8.88	Diethanolamine	−0.025
9.05	4-Phenolsulphonic acid	
9.06	2-Aminoethylsulphonic acid	−0.022
9.10	Alanine (pK_2)	

APPENDIX III: (*Cont.*)

pK_a (25°C)	Substance	dpK_a/dT
9.11	4-Aminopyridine	−0.028
9.18	Histidine	
9.21	Serine	−0.025
9.23	Boric acid	−0.008
9.24 §	Tetraethylethylene diamine (pK_1)	
9.25	Ammonia	−0.031
9.41	Pyrophosphoric acid (pK_4)	
9.50	Ethanolamine	−0.029
9.54	Ephedrine	−0.022
9.55 (20°C)*	CHES (3-(Cyclohexylamino)ethanesulphonic acid)	
9.55 §	Dimethylaminoethylamine (pK_1)	
9.66	Hydroxyproline	−0.023
9.69	2-Amino-2-methyl-1-propanol	−0.032
9.74	Leucine	−0.026
9.78	Glycine (pK_2)	−0.025
9.80	Trimethylamine	−0.022
9.81	Piperazine (pK_1)	
9.87	Alanine (pK_2)	−0.027
9.93	Ethylenediamine (pK_1)	−0.029
9.96	1-Aminopropan-3-ol	
10.00	Aspartic acid (pK_3)	−0.022
10.00	Phenol	
10.24	β-Alanine (pK_2)	−0.028

APPENDIX III: *(Cont.)*

pK_a (25°C)	Substance	dpK_a/dT
10.33	Carbonic acid (pK_2)	−0.009
10.40 (20°C)*	CAPS (3-(Cyclohexylamino)propanesulphonic acid)	
10.56	γ-Aminobutyric acid	−0.030
10.57	*n*-Propylamine	−0.032
10.62	Methylamine	−0.032
10.63	Ethylamine	−0.032
10.64	*n*-Butylamine	−0.032
10.64	sec-Butylamine	
10.72	Triethylamine	−0.033
10.77	Dimethylamine	−0.033
10.93	Diethylamine	−0.034
10.93	Hexamethylenediamine (pK_1)	−0.034
11.12	Piperidine	−0.031
12.33	Phosphoric acid (pK_3)	−0.026
13.6	Guanidine	

* 'Apparent' or 'practical' constant for 0.1M solution
† 'Apparent' or 'practical' constant for 0.05M solution
‡ 'Apparent' constant, I = 0.025
§ 'Apparent' constant, I = 0.1
‖ pK_1' = 4.36 (c = 0.01M)

Appendix IV.

The Henderson—Hasselbalch equation.

The equation, given originally for the CO_2/H_2CO_3 system (Henderson, 1908; Hasselbalch, 1916), can be derived as follows:

Consider a solution prepared by adding C_a moles of a weak acid, HA, to C_b moles of the salt, NaA, of its conjugate base, A^-, in a volume of 1 litre. Let h = antilog $(-pH)$, measured by the glass electrode, be assumed to be the hydrogen ion activity in the final solution.

Five species in the system, namely HA, A^-, Na^+, H^+ and OH^-, are related by

$$h(A^-)/(HA) = K_a \tag{10.1}$$

$$h(OH^-) = K_w \tag{10.2}$$

$$[HA] + [A^-] = C_a + C_b \tag{10.3}$$

$$[Na^+] = C_b \tag{10.4}$$

$$[H^+] + [Na^+] = [OH^-] + [A^-] \tag{10.5}$$

Equation 10.3 is a material balance relation, and Equation 10.5 is required for electroneutrality. The parentheses denote activities and square brackets represent concentrations. Introducing activity coefficients and rearranging, Equations 10.1 and 10.2 become

$$h[A^-]/[HA] = K_a f_{HA}/f_A = K_a' \tag{10.6}$$

$$h[OH^-] = K_w f_{OH} \tag{10.7}$$

Also $\qquad\qquad [H^+] = h/f_H \tag{10.8}$

Taking logarithms and rearranging, Equation 10.6 can be written

$$pH = pK_a' + \log [A^-]/[HA] \tag{10.9}$$

This expression is correct, but it is often *wrongly* assumed that [HA] is always equal to C_a and that $[A^-]$ is equal to C_b. This misconception arises from the assumption that the ionization of HA is completely repressed in the presence of A^-. The correct relation is derived below.

Substitution of Equation 10.4 into Equation 10.5 gives

$$[A^-] = C_b + h/f_H - K_w f_{OH}/h \qquad (10.10)$$

and hence, from Equation 10.3,

$$[HA] = C_a - h/f_H + K_w f_{OH}/h \qquad (10.11)$$

$$\therefore K_a' = h \frac{(C_b + h/f_H - K_w f_{OH}/h)}{(C_a - h/f_H + K_w f_{OH}/h)} \qquad (10.12)$$

that is, $pH = pK_a' + \log \dfrac{(C_b + h/f_H - K_w f_{OH}/h)}{(C_a - h/f_H + K_w f_{OH}/h)}$ (10.13)

This reduces to the commonly used approximate form of the equation only if $[H^+]$ and $[OH^-]$ are small compared to C_a or C_b.

Below pH 7, $K_w f_{OH}/h$ is negligible, so that for acid solutions Equation 10.13 can be written

$$pH = pK_a' + \log \frac{(C_b + h/f_H)}{(C_a - h/f_H)} \qquad (10.14)$$

Rearranging as a quadratic in h, and solving, gives

$$[H^+] = h/f_H = \{((C_b + K_a' f_H)^2 + 4 C_a K_a'/f_H)^{\frac{1}{2}}$$
$$- (C_b + K_a'/f_H)\}/2 \qquad (10.15)$$

Above pH 7, h/f_H is negligible compared to C_a and C_b, so that Equation 10.13 becomes

$$pH = pK_a' + \log \frac{(C_b - K_w f_{OH}/h)}{(C_a + K_w f_{OH}/h)} \qquad (10.15)$$

$$= pK_a' + \log \frac{(C_b - (OH^-)f_{OH})}{(C_a + (OH^-)f_{OH})} \qquad (10.16)$$

Solving for h gives

$$h = (K_a' C_a + K_w f_{OH}) + \{(K_a' C_a + K_w f_{OH})^2 + 4 K_a' K_w f_{OH} C_b)^{1/2}\}/2 C_b$$
$$(10.17)$$

Similar relations can be derived for other charge types of acid/base pairs.

When the hydrogen or hydroxyl concentrations are less than one hundredth of the concentrations of the buffer acidic or basic species the error introduced by neglecting the terms in $[H^+]$ and $[OH^-]$ is less than 0.009 pH unit. If the ration reaches one tenth the error becomes 0.09 pH unit. Thus, the simple equation is suitable as follows:

C_a or C_b	pH range
0.005M	4.3 – 9.7
0.01	4 – 10
0.05	3.3 – 10.7
0.1	3 – 11

Outside these ranges, the extended equation should be used.

References

Albert (1968), 'Selective Toxicity', 4th Ed., London: Methuen.
Albert and Serjeant (1971), 'The Determination of Ionization Constants', 2nd Ed., London: Chapman and Hall.
Alfenaar and de Ligny (1967), *Rec. Trav. chim. Pays-Bas*, **86**, 1185.
Alner, Greczek and Smeeth (1967), *J. Chem. Soc.*, **(A)**, 1205.
Aronsson and Grönwall (1957), *Scand. J. Clin. Lab. Invest.*, **9**, 338.
Ashton (1957), *Nature, London*, **180**, 917.

Bates (1962), *J. Res. Nat. Bureau Standards*, **66A**, 179.
Bates (1964), 'Determination of pH, Theory and Practice', New York: Wiley.
Bates (1969), *Anal. Chem.*, **41**, 283.
Bates and Bower (1956), *Anal. Chem.*, **28**, 1322.
Bates and Guggenheim (1960), *Pure and Appl. Chem.*, **1**, 163.
Bates, Paabo and Robinson (1963), *J. Phys. Chem.*, **67**, 1833.
Bates, Pinching and Smith (1950), *J. Res. Nat. Bureau Standards*, **45**, 418.
Bax, de Ligny and Remijnse (1973), *Rec. Trav. chim. Pays-Bas*, **92**, 374; and earlier papers.
Best and Samuel (1936), *Ann. Appl. Biol.*, **23**, 509.
Bloemendal (1963), 'Zone Electrophoresis in Blocks and Columns', Amsterdam: Elsevier, pp. 24, 68.
Blombäck, Blombäck, Edman and Hessel (1966), *Biochim. Biophys. Acta*, **115**, 371.
Boman and Westlund (1956), *Arch. Biochem. Biophys.*, **64**, 217.
Bower and Bates (1955), *J. Res. Nat. Bureau Standards*, **55**, 197.
Braunitzer and Schrank (1970), *Hoppe-Seyler's Zeits. physiol. Chem.*, **351**, 417.
Breslow (1972), *Chem. Soc. Reviews*, **1**, 553.
Britton and Robinson (1931), *J. Chem. Soc.*, 458.
Britton and Robinson (1931a), *J. Chem. Soc.*, 1456.
Britton and Welford (1937), *J. Chem. Soc.*, 1848.
Broser and Fleischhauer (1970), *Zeits. Naturforsch.*, **25B**, 1389.
Buchanan and Hamann (1953), *Trans. Faraday Soc.*, **49**, 1425.

Carmody (1961), *J. Chem. Ed.*, **38**, 559.
Ceccarini and Eagle (1971), *Proc. Nat. Acad. Sci. U.S.*, **68**, 229.
Chaberek and Martell (1959), 'Organic Sequestering Agents', New York: Wiley.
Clark and Lubs (1916), *J. Biol. Chem.*, **25**, 479.
Coch Frugoni (1957), *Gazz. Chim. Ital.*, **87**, 403.

Covington, Paabo, Robinson and Bates (1968), *Anal. Chem.*, **40**, 700.
Covington, Robinson and Bates (1966), *J. Phys. Chem.*, **70**, 3820.
Cutie and Sciarrone (1969), *J. Pharm. Sci.*, **58**, 990.

Davies (1938), *J. Chem. Soc.*, 2093.
Davies (1959), *Analyst*, **84**, 248.
Davies and Hoyle (1953), *J. Chem. Soc.*, 4134.
Davies and Hoyle (1955), *J. Chem. Soc.*, 1038.
Davis and Mingioli (1950), *J. Bact.*, **60**, 17.
Dawson, Elliott, Elliott and Jones (1969), 'Data for Biochemical Research', 2nd Ed., Oxford: Oxford Univ. Press.
Delory and King (1945), *Biochem. J.*, **39**, 245.
Douheret (1967), *Bull. Soc. chim. France*, 1412.
Douheret (1968), *Bull Soc. chim. France*, 3122.
Durst (ed.), *Nat. Bureau Standards Technical Note*, **543**, 10.
Durst and Staples (1972), *Clin. Chem.*, **18**, 206.

Eagle (1959), *Science*, **130**, 432.
Eagle (1971), *Science*, **174**, 500.
Edman (1950), *Acta Chem. Scand.*, **4**, 283.
Edman and Begg (1967), *Europ. J. Biochem.*, **1**, 80.
Ellis (1961), *Nature, London*, **191**, 1099.
Elving, Markowitz and Rosenthal (1956), *Anal. Chem.*, **28**, 1179.

Fels (1904), *Zeits. Elektrochem.*, **10**, 208.
Fenn (1973), Personal communication.
Fernbach and Hubert (1900), *Compt. rend.*, **131**, 293.
Florence and Dempsey (1973), Personal communication.
Fossum, Markunas and Riddick (1951), *Anal. Chem.*, **23**, 491.
Frant and Ross (1968), *Anal Chem.*, **40**, 1169.

Gelsema, de Ligny and Visserman (1965), *Rec. Trav. chim. Pays-Bas*, **84**, 1129.
Gensch (1967), *Arzneim.-forsch.*, **17**, 802.
German and Vogel (1937), *Analyst*, **62**, 271.
Glasoe and Long (1960), *J. Phys. Chem.*, **64**, 188.
Gomori (1946), *Proc. Soc. Exp. Biol. Med.*, **62**, 33.
Gomori (1955), *Methods in Enzymology*, **1**, 141.
Good, Winget, Winter, Connolly, Izawa and Singh (1966), *Biochem.*, **5**, 467.
Gottschalk (1959), *Zeit. anal. Chem.*, **167**, 342.
Gottschalk (1961), *Zeit. anal. Chem.*, **183**, 420.
Gregory and Sajdera (1970), *Science*, **169**, 97.

Hamann (1963), *J. Phys. Chem.*, **67**, 2233.
Hamann and Strauss (1955), *Trans. Faraday Soc.*, **51**, 1684.
Hasselbalch (1916), *Biochem. Zeits.*, **78**, 112.

Havas, Kaszas and Varsanyi (1971), *Proc. 2nd Conf. Appl. Phys. Chem.*, 1, 529.
Henderson (1908), *Amer. J. Physiol.*, 21, 169.
Hetzer, Robinson and Bates (1968), *Anal. Chem.*, 40, 634.
Hitchcock and Taylor (1938), *J. Amer. Chem. Soc.*, 60, 2710.

Ingri, Kakolowicz, Sillén and Warnqvist (1967), *Talanta*, 14, 1261.
Irving and Cox (1958), *Analyst*, 83, 526.
Israel (1949), *Proc. Roy. Austral. Chem. Inst.*, 16, 97.

Jermyn (1967), *Austral. J. Chem.*, 20, 183.

Kalckar (1947), *J. Biol. Chem.*, 167, 429.
Keyworth and Hahn (1958), *Talanta*, 1, 41.
Kolthoff (1925), *J. Biol. Chem.*, 63, 135.
Kolthoff and Fischgold (1932), 'Säure-Basen Indicatoren', Berlin: Springer.
Kortüm, Vogel and Andrussow (1961), 'Dissociation Constants of Organic Acids in Aqueous Solution', London: Butterworths.
Krebs (1951), *Biochem. J.*, 48, 349.
Krebs and Henseleit (1932), *Hoppe-Seyler's Zeits. physiol. Chem.*, 210,33.
Kuntz, Loach and Calvin (1964), *Biophys. J.*, 4, 227.

Leonis (1948), *Compt. Rend. Trav. Lab. Carlsberg. Ser. Chim.*, 26, 357.
Lerch and Stegemann (1969), *Anal. Biochem.*, 29, 76.
Lesquibe and Reynaud (1970), *Compt. rend.*, C, 271, 1193.
Lewis (1966), *Anal. Biochem.*, 14, 495.
de Ligny and Alfenaar (1967), *Rec. Trav. chim. Pays-Bas*, 86, 1182.
de Ligny, Luykx, Rehbach and Wieneke (1960), *Rec. Trav. chim. Pays-Bas*, 79, 713.
Lindberg and Swan (1960), *Acta Chem. Scand.*, 14, 1043.
Long (1961), 'Biochemists' Handbook', London: Spon, p. 30.

McIlvaine (1921), *J. Biol. Chem.*, 49, 183.
MacInnes, Belcher and Shedlovsky (1938), *J. Amer. Chem. Soc.*, 60, 1094.
Mallette (1967), *J. Bact.*, 94, 283.
Maravalhas (1969), *J. Chromatog.*, 44, 617.
Matsumura, Takaoka and Katsuta (1968), *Exp. Cell Res.*, 53, 337.
Mellon, Acree, Avery and Slagle (1921), *J. Infect. Diseases*, 29, 7.
Mertz and Owen (1940), *Proc. Soc. Exp. Biol. Med.*, 43, 204.
Michaelis (1930), *J. Biol. Chem.*, 87, 33.
Miller and Golder (1950), *Arch. Biochem.*, 29, 420.

Nightingale (1958), *Anal. Chem.*, 30, 267.

Ong, Robinson and Bates (1964), *Anal. Chem.*, **36**, 1971.
Orr, Blakley and Panagou (1972), *Anal. Biochem.*, **45**, 68.

Paabo and Bates (1969), *Anal. Chem.*, **41** 283.
Paabo, Bates and Robinson (1963), *J. Res. Nat. Bureau Standards*, **67A**, 573.
Paabo, Robinson and Bates (1965), *J. Amer. Chem. Soc.*, **87**, 415.
Paabo, Robinson and Bates (1966), *Anal. Chem.*, **38**, 1573.
Palitzsch (1915), *Biochem. Zeits.*, **70**, 333.
Peeters (1959), *Advances in Clin. Chem.*, **2**, 1.
Perrin (1963), *Austral. J. Chem.*, **16**, 572.
Perrin (1964), *Austral. J. Chem.*, **17**, 484.
Perrin (1965), 'Dissociation Constants of Organic Bases in Aqueous Solution', London: Butterworths.
Perrin (1969), *Pure and Appl. Chem.*, **20**, 133.
Perrin (1970), 'Masking and Demasking of Chemical Reactions', New York: Wiley-Interscience.
Perrin (1972), 'Dissociation Constants of Organic Bases in Aqueous Solution. Supplement', London: Butterworths.
Perrin, Armarego and Perrin (1966), 'Purification of Laboratory Chemicals', Oxford: Pergamon Press.
Perrin and Sayce (1967), *Talanta*, **14**, 833.
Perry, Stedman and Hansen (1968), *J. Chromatog.*, **38**, 460.
Peters, Berridge, Cummings and Lin (1968), *Anal. Biochem.*, **23**, 459.
Pilz and Johann (1966), *Zeits. anal. Chem.*, **215**, 105.
Plumel (1948), *Bull. Soc. Chim. biol.*, **30**, 129.
Popa, Enea and Luca (1965), *Zeits. physik. Chem. (Leipzig)*, **230**, 271.
Popa, Luca and Enea (1965a), *Zeits. physik. Chem. (Leipzig)*, **230**, 262.
Poulik (1957), *Nature, London*, **180**, 1477.
Prideaux and Ward (1924), *J. Chem. Soc.*, **125**, 426.

Rauflaub (1956), 'Methods of Biochemical Analysis', (ed. Glick), **3**, p. 301, New York: Interscience.
Remizov (1960), *Biokhimia*, **25**, 323.
Reynaud (1969), *Compt. rend., C*, **269**, 777.
Ringbom (1963), 'Complexation in Analytical Chemistry', New York: Interscience.
Robertson and Boyer (1956), *Arch. Biochem. Biophys.*, **62**, 396.

Scatchard and Prentiss (1934), *J. Amer. Chem. Soc.*, **56**, 2314.
Schwabe, Graichen and Spiethoff (1959), *Zeits. physik. Chem. (Frankfurt)*, **20**, 68.
Schwarzenbach (1957), 'Complexometric Titrations', New York: Interscience.
Schwarzenbach, Anderegg, Schneider and Senn (1955), *Helv. Chim. Acta*, **38**, 1147.
Semenza, Landucci and Mülhaupt (1962), *Helv. Chim. Acta*, **45**, 2306.

Semple, Mattock and Uncles (1962), *J. Biol. Chem.*, **237**, 963.
Shipman (1969), *Proc. Soc. Exp. Biol. Med.*, **130**, 305.
Sillén (1967), *Chem. in Britain*, **3**, 291.
Sillén and Martell (1964), 'Stability Constants of Metal-ion Complexes', London: Chem. Soc. Special Publ. No. 17.
Sillén and Martell (1971), 'Stability Constants of Metal-ion Complexes. Supplement', London: Chem. Soc. Special Publ. No. 25.
Simon (1964), *Angew. Chem., Internat. Edn.*, **3**, 661; and references therein.
Smith and Smith (1949), *Biol. Bull.*, **96**, 233.
Smithies (1959), *Adv. Protein Chem.*, **14**, 68.
Smithies (1962), *Arch. Biochem. Biophys. Suppl.*, **1**, 125.
Smits (1947), *Biochim. Biophys. Acta*, **1**, 280.
Sørensen (1909), *Biochem. Zeits.*, **21**, 131.
Sørensen (1912), *Ergebn. Physiol.*, **12**, 393.
Spinner and Petersen (1961), *Scand. J. Clin. and Lab. Invest.*, **13**, 1.
Stafford, Watson and Rand (1955), *Biochim. Biophys. Acta*, **18**, 318.
Staples and Bates (1969), *J. Res. Nat. Bureau Standards*, **73(A)**, 37.
Stinson and Spencer (1968), *Canad. J. Biochem.*, **46**, 43.
Strickland and Anderson (1966), *Anal. Chem.*, **38**, 980.
Svensson (1962), *Acta Chem. Scand.*, **16**, 456.
Swartz and Wilson (1971), *Anal. Biochem.*, **40**, 392.

Tanaka (1963), *Anal. Chim. Acta*, **29**, 193.
Temple (1929), *J. Amer. Chem. Soc.*, **51**, 1754.
Teorell and Stenhagen (1938), *Biochem. Zeits.*, **299**, 416.
Thies and Kallinich (1953), *Biochem. Zeits.*, **324**, 485.
Tobie and Ayres (1945), *J. Bact.*, **50**, 333.
Tuddenham and Anderson (1950), *Anal. Chem.*, **22**, 1146.

Valensi (1972), *Pure and Appl. Chem.*, **31**, 547.
Van Slyke (1922), *J. Biol. Chem.*, **52**, 525.
Van Veen, Hoefnagel and Wepster (1971), *Rec. Trav. chim. Pays-Bas*, **90**, 289.

Whiting (1966), *Chem. and Ind. (London)*, 1030.
Wolf (1973), *Experientia*, **29**, 241.
Woodhead, Paabo, Robinson and Bates (1965), *Anal. Chem.*, **37**, 1291.

Yamazaki and Tolbert (1970), *Biochim. Biophys. Acta*, **197**, 90.

Index